THE CHANGING GEOGRAPHY OF AFRICA AND THE MIDDLE EAST

This book is about the geography of change in Africa and the Middle East. For nearly all of these countries the doubling time for the population is less than thirty years, in the case of Kenya it is only seventeen years. Rates of urbanization are high, and increasing education and the media are changing the political awareness and expectations of these populations. These countries have to struggle simultaneously with the problems of feeding themselves, finding new employment opportunities, and servicing overseas debt, and now also the increasing impact of the AIDS epidemic. Tensions have spawned conflicting political ideologies. In Southern Africa and the Horn of Africa the people have had to contend with bitter and long-lasting wars. Increasingly these nations are facing environmental problems, brought about by over-use of fragile soils, or by increasing frequencies of drought. Some are fortunate enough to have valuable resources, such as oil, that have helped them in the short or medium term at least, others have nothing.

In this book experts at, or associated with, the Department of Geography of the School of Oriental and African Studies give students of Geography and allied subjects an up-to-date understanding of these changes. The first introductory chapter surveys Africa and the Middle East as these regions were understood in the mid-1960s. This was a time of hope – a new dawn for many of the states which had just become independent. In the following chapters specialists consider what has happened since in the regions of their own expert knowledge. Many of these countries, even those endowed with significant natural resources, have since changed in unanticipated ways, as the tensions have grown within plural societies, each needing to develop new institutions of their own while at the same time agreeing on the strategy which would best meet their development aims and the people's expectations. The last chapter summarizes and reviews the changes that have taken place, showing how differently these areas are seen now compared with twenty-five years ago.

Graham P. Chapman and **Kathleen M. Baker** are, repectively, Professor and Lecturer in the Department of Geography, School of Oriental and African Studies, University of London.

D0162126

THE CHANGING GEOGRAPHY OF AFRICA AND THE MIDDLE EAST

Edited by
Graham P. Chapman and
Kathleen M. Baker

FOR THE DEPARTMENT OF GEOGRAPHY AT SOAS

London and New York

First published 1992
by Routledge
11 New Fetter Lane, London EC4P 4EE

Simultaneously published in the USA and Canada
by Routledge
a division of Routledge, Chapman and Hall, Inc.
29 West 35th Street, New York, NY 10001

© 1992 Graham P. Chapman and Kathleen M. Baker
Typeset in Garamond by J&L Composition Ltd, Filey,
North Yorkshire
Printed and bound in Great Britain by
Biddles Ltd, Guildford and King's Lynn

British Library Cataloguing in Publication Data
A catalogue record for this book is available from the British
Library.
ISBN 0–415–05709–4 (HB)
ISBN 0–415–05710–8 (PB)

Library of Congress Cataloging in Publication Data
The Changing geography of Africa and the Middle East /
edited by Graham P. Chapman and Kathleen M. Baker for the
Geography Department at SOAS.
Includes bibliographical references and index.
ISBN 0–415–05709–4 (HB). – ISBN 0–415–05710–8 (PB)
1. Africa–Geography. 2. Africa–Politics and government–
1960–. 3. Africa–Economic conditions–1960–. 4. Middle
East–Geography. 5. Middle East–Politics and government–
1945–. 6. Middle East–Economic conditions–1945–.
I. Chapman, Graham P. II. Baker, Kathleen M., 1950–.
III. University of London. School of Oriental and African
Studies. Dept. of Geography.
DT6.7.C47   1992
960.3'2–dc20
91–44795 CIP

CONTENTS

v

FIGURES

TABLES

CONTRIBUTORS

J. Anthony Allan is Professor of Geography at the School of Oriental and African Studies

Felicité Awassi Atsimadja is a PhD student at the School of Oriental and African Studies

Dr Kathleen Baker is Lecturer in Geography at the School of Oriental and African Studies

Graham Chapman is Professor of Geography at the School of Oriental and African Studies

Dr Anthony O'Connor is Reader in Geography at University College London and a member of the Centre for African Studies at the School of Oriental and African Studies

St John B. Gould is a PhD student at the School of Oriental and African Studies

George Joffé is with the Economist Intelligence Unit, and a member of the Centre for Middle East Studies at the School of Oriental and African Studies

Keith McLachlan is Professor of Geography at the School of Oriental and African Studies

Dr Deborah Potts is Lecturer in Geography at the School of Oriental and African Studies

DISCLAIMER FROM THE AUTHORS AND THE PUBLISHERS

Whilst we have made every effort to ensure that all maps, diagrams, figures and data are as accurate as possible, we do not claim to represent the official position of any government or other agency with regard to the position of any borders, to the names of any places, nor to statistics.

PREFACE

The undergraduate programme in the Department of Geography at the School of Oriental and African Studies started in 1965, just over twenty-five years ago as we write. The department decided to celebrate this quarter of a century by writing a Geography of Change, on the areas of the world to which our study and teaching is committed. This comprehensive review is being published in this and the companion volume on *The Changing Geography of Asia*.

It has of course been a daunting exercise, particularly to compress into books of this length some sense of the detailed knowledge of their areas which each expert in the department embraces. We have tried to do so at a level which is accessible to undergraduates and to serious A-level students, and we believe that we have succeeded in our aim – although obviously it is for the reader to decide.

Higher education in the UK is passing through a period of great stress. Resources for research and publication are being continually eroded, and it is notable that in the UK as a whole the study of geography at school and university levels has retreated from the broader overseas perspective it used to have, sometimes into esoteric theory building, sometimes into empirical studies which stress the UK or other developed nations. Clearly though, the world is becoming more and more interdependent, no matter whether in trade or defence, finance or in recognition of global climatic change. We need now, more than ever, a well-educated and well-informed citizenry, whose understanding of the wider world we live in can inform the conduct of their lives, and who can contribute to public opinion formation on vital contemporary issues as well.

The knowledge that such opinions are founded on cannot be fixed. The world is changing so rapidly, that new knowledge generation at higher rates is even more important now than in the past. This means that geography in the UK must revitalize its traditional world-wide vision, and must be given the resources to complete its mission. We hope that this book will stimulate some awareness of the magnitude and speed of change, and the seriousness of the problems of understanding now and in the future. Perhaps, with luck, we will stimulate some students to join the woefully small band of experts who have committed their careers to this end.

ACKNOWLEDGEMENT

The authors would like to thank Mrs Catherine Lawrence for the good-natured way in which she produced such high-quality maps, often at short notice.

1

INTRODUCTION

Independence: promise at the new dawning

Graham Chapman and Kathleen Baker

AFRICA SOUTH OF THE SAHARA

As recently as 1955, no present African state existed as an independent nation state, with the exception of Liberia, Egypt, Ethiopia and South Africa, but by 1975, two decades later, the entire continent was independent and free from colonial domination – in theory, if not in practice. Most states in Africa gained their independence in or around 1960. It was only for the Portuguese states that the struggle for independence continued with long and bloody wars being fought in Angola, Mozambique which became independent in 1975 and in one of the continent's tiniest states, Guinea-Bissau, which became independent in 1974 (see Figure 1.1).

The intention in this introduction is to go back some twenty-five or thirty years to a time when the majority of African nations were either newly independent, or soon to be so, and to consider just some of the different perceptions of independence, and the expectations of what freedom from colonial rule would bring. The views of Africans towards their independence, of the colonizers, and of academics, are among those considered, and it is apparent that independence promised very different things to different people.

To Africans, bringing colonial domination to an end was understandably what independence was all about. European rulers of Africa had long treated Africans as inferior both as individuals and as a race, and had perceived them as backward in terms of their development (Perham 1961b: 852). That Africans were simply not ready for self-government was the response patronizingly meted out to ever more frequent requests for self-government. The

1

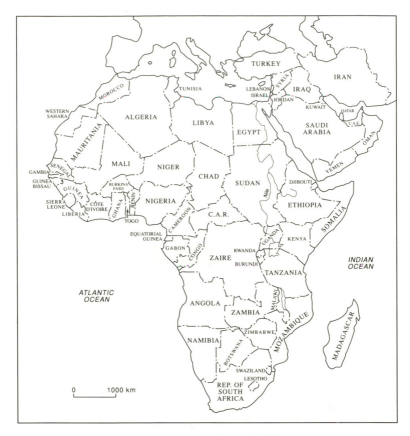

Figure 1.1   Africa and the Middle East

bitterness of feeling towards the European in Africa was not hidden by many Africans, including Kwame Nkrumah, who gave weight to the view that colonialism was rooted in self-interest, not in altruism. Colonial governments, he argued, used the colonies as a source of raw materials and cheap labour, and as a dumping ground to market their produce at exorbitant prices. He also argued that while colonial involvement in Africa was fundamentally economic, the solution to this 'problem' was political, with political independence being an essential step towards 'economic emancipation' (Minogue and Molloy 1974: 21).

The preoccupation with politics by all concerned with African independence, and the lack of attention paid to Africa's economic

2

future in the late 1950s and early 1960s, is lamented by Frankel (1961). However, Oginga Odinga, one time Vice President of Kenya, goes some way to explaining why political issues in the independence struggle so overshadowed economic matters: colonial administrators had advised Africans that political power would be gained only once they had demonstrated their capability of achieving economic advancement. Speaking of Kenya, Oginga Odinga argues that such advice was merely to give colonial administrations more time. Wherever Africans turned in order to achieve economic advance, government-made obstacles loomed to prevent them. Comparable attitudes by colonial administrators in other parts of the continent thus led Africans to conclude that the struggle for political power was fundamental to any advance at all, and the focus on economic development could come later (Minogue and Molloy 1974: 39). In the movement towards Zambia's independence, Kenneth Kaunda stressed that the success of the colonials in Africa was the result of their superior political organization, and that effective political organization was fundamental if Africa was to be freed from colonial domination. It is therefore not surprising that gaining political independence seemed to be all that mattered.

Not all Africans felt so strongly against the Europeans. In the mid-1950s, Patrice Lumumba still believed that the colonial government had the interests of both the Congolese people and the Belgians at heart in the development of the Congo. He asked, 'To whom do we owe our liberation from that odious trade practised by the bloodthirsty Arabs and their allies, those inhuman brigands who ravished the country?' With the assistance of the Africans, he stated that the Belgians were 'able to rout the enemy, to eradicate disease, to teach us and to eliminate certain barbarous practices from our customs, thus restoring our human dignity, and turning us into free, happy, vigorous, civilised men' (Minogue and Molloy 1974: 42–3).

Such positive views towards the Europeans changed as Africans perceived that independence was being placed beyond their reach. Tension rose, particularly in Africa's urban areas, and it was here, where Africans came into closest contact with Europeans, that the independence movements were strongest. The greater part of rural Africa however, remained relatively unchanged by colonialism – remote from colonials, if not from many of the economic changes wrought by colonialism. But some rural areas were not too remote and had been subjected to the demands of colonial governments for taxation, and for forced labour among other things. In these areas too feelings against colonials ran high. Thus it was that political independence or, getting rid of European rule, was the dominant issue to many Africans, and thoughts of economic and social implications were of secondary importance until Africa could be freed from European domination.

In contrast to numerous occasions in history when one ethnic group or country has exercised power over another, in this case, the ending of colonialism was accompanied by a moral commitment by the colonizers to assist the development of former colonies. A large measure of this was of course self-interest, to counter the perceived threat from the growth of communism. Such was the concern that in 1961, the House of Lords proposed that 'Britain should "ensure" African independence by teaching Africans not to be Communists' (Fordham 1974: 264). Western liberalism which pledged support to assist with the development of former colonies was clearly shocked by the rapid growth of anti-colonial feeling, and the bitterness of it. Such bitter feelings were not new, but improved communications throughout the world brought them to the attention of a wider audience than ever before, an audience formerly unaware of such issues, and still surprised by the strength of feeling against former colonial powers – even after independence.

The euphoria of the Africans, and their high expectations of what independence would bring, were not matched by those who had had long experience of working and living in Africa. First, there were doubts about the political viability of some states in Africa. The new African countries had never before survived as independent nation states, and the boundaries drawn by colonial administrators had brought together groups in racial and ethnic conflict – Nigeria, Sudan, Uganda, Ethiopia and Somalia among others. Nigeria for example was divided between the predominantly Muslim north, the Yoruba west and the Ibo east. Speaking in 1961 at the time of independence, the former governor Sir James Robertson stated that forty-seven years was a very short time in which to weld together the many different people of Nigeria into a nation, though he was of the opinion that political and economic forces which bound Nigeria together were greater than those which would divide it (Robertson 1961).

Second, there was concern over the economic future of the continent. The economic viability of nations as small as The Gambia was in question, and there was the possibility that these would be open to the threat of absorption by larger neighbouring nations. An equally difficult future was foreseen for some of the Sahel states such as Mali, Niger and Chad. These covered vast areas, but included relatively few permanent residents within their boundaries. Speaking again of Nigeria with which he was understandably most concerned, Robertson was cautiously optimistic that the new state would 'be all right' largely due to the diverse nature of its economy. Rich produce from the forested south including cocoa, palm products, rubber and copra were all at the time valuable earners of foreign exchange, and added to these was produce from the north – groundnuts, cotton and animal products. Furthermore, the inclusion of varied ecological environments within the boundaries of Nigeria and other nations such as Côte d'Ivoire, Cameroon and Ghana offered valuable protection against loss of food crops through drought or disease. There was, however, concern for nations which depended heavily on a single crop for foreign exchange – cocoa in the case of Ghana was mentioned by Robertson, but many others, particularly the Sahelian states, were even more vulnerable. So much depended on the stability of price and also of climate – and these were unpredictable.

A third cause of pessimism about the future was that in the years leading up to independence many potential problems began to 'harden out'. Colonial governments had successfully contained much of the ethnic conflict, but with independence many of these problems surfaced rapidly to confront new African governments. It was an awareness of the potentially divisive and destructive effects of these issues that led Sir Charles Arden-Clarke, the last governor of Ghana, to comment rather bluntly that the major threats to Africa's independent future would be tribalism, poverty and ignorance. Arden-Clarke was clearly quite right about tribalism and poverty . With regard to 'ignorance' we must take issue with such a patronizing term, as it is this view of the 'ignorant African' which has done so much to undermine Africa's potential for development. Failure to understand and respect African social and political organization as well as African use of the land was one of the worse features that often accompanied colonialism. For centuries locally increasing population pressure on the land in Africa has led to increasing pressure on environmental resources. However, by skilful use of management techniques, economies were sustainable. But although, even in colonial times, the threat of environmental degradation was apparent, the long-term impact of such threats was never taken seriously, so strong was the urge to establish economically viable forms of agriculture – regardless of ecological soundness.

The changes that took place in Africa some twenty-five to thirty years ago were also of immense interest to academics of all disciplines. Social scientists for example were deeply interested in, among other things, the nature of government in the newly independent states. There was also much academic interest in the development strategies produced by nations with different political ideologies. They ranged from Tanzania's radical attempt to create a completely new and socialist path to development, to the very much more capitalist approaches of Kenya and Côte d'Ivoire.

Regarding the nature of future change, academics and other interested individuals tended to fall into two 'camps': first, and by far the larger, supported the view that Africa would develop acording to the Western Model. In the second camp were those, fewer in number, who believed that development of Africa was destined to be quite different from that in the West. Opinion as to the nature of development under these circumstances was far less clear. Among the members of the former group was Immanuel Wallerstein who predicted that development would occur along Western lines, although he argued development could take a very long time. Writing in 1964 of the Côte d'Ivoire and Ghana, Wallerstein (1964) stated that 'The forms of the new society are not unknown'. He may well be proved correct ultimately, but at present there is little evidence that development has occurred along predicted lines.

It would appear that many new African governments also anticipated that post-independence development would follow development patterns in the West, but over a much shorter time-scale. For Ghana, for example, the way ahead was to be through industrialization, though it was at the expense of the rural sector. Profits from a relatively prosperous agricultural sector were invested in allegedly more progressive sectors, all in the name of creating a modern economy. Nigeria, convinced in the early days of independence that the rural sector should be developed, later turned away from its traditional farmers and allocated considerable funds to the establishment of large-scale, capital intensive, Western style projects. Some of these had even been given the go-ahead by experts of international organizations such as FAO, USAID and the Commonwealth Development Corporation (Andrae and Beckman 1985:

77–87). Output from Nigeria's large-scale agricultural projects would theoretically more than satisfy national demand, would provide necessary resources for industry, and would reduce national dependence on traditional farmers considered to be outmoded.

René Dumont (1962; 1988) has always been one of the strongest opponents of the view that development should follow Western lines, not least because Africa south of the Sahara is environmentally a very different place from Europe and North America. He argued for development backed by appropriate forms of education for Africans, and for sound rural development where agricultural produce could be processed locally in order to build up the strength of local economies. Essentially, he argued for a greater input by Africans, and the search – by both Africans and Westerners – for appropriate solutions to specific problems in Africa.

Dumont argued that with considerable assistance from the West and with greater African effort, conditions in Africa could improve. However, it all came down to identifying assets which could be developed. Hirschmann (1965) argued that what was seen as an obstacle in one context could be an advantage or an asset in others, and equally, an advantage in one context may not be so in another. In Africa, where for too long domination by foreign powers has bred attitudes of white superiority and transmitted them across the world, it was important that African leaders and people in other developing nations took stock and reappraised their national development potential. Trying to re-create development along Western lines was not the way ahead. These views were supported by Davis Fogg (1965) and others with long experience of Africa who argued that aspects of traditional society, not least traditional agriculture, should be made the cornerstone of future development.

Writing in 1986, Fieldhouse remarks that one of the most astonishing features of post-1950 African history was the general expectation that independence would lead to rapid economic growth and affluence. He argues that there was no historical precedent for this, as the wealth of Europe and North America had been accumulated over a considerable length of time. But there are those who wonder, now, if Taiwan or South Korea, have made the leap, why not Africa?

Africa north of the Sahara was always seen by Europeans in a very different light from darkest sub-Saharan Africa. The states of this region, with the exception of Sudan and Mauritania, all fringe the Mediterranean sea at least at one point. They have not only been the cradle of their own ancient civilizations but also throughout much of their history been heavily involved with the civilizations that have emerged on the North Mediterranean coast; but contact and involvement does not necessarily mean a convergence of culture. North Africa is distinctive from Europe in that its population predominantly uses non-Latin languages, the commonest being various dialects of Arabic, and in that its populations are overwhelmingly Muslim.

From the early or late nineteenth century European colonial powers began to dominate the area. The establishment of a French colonial state in Algeria was followed by major waves of French settlement, which lead ultimately to an unsuccessful proposal to incorporate Algeria as part of metropolitan France, and a bitter war against guerrillas fighting for independence. The Italians briefly colonized Libya and a part of Ethiopia. But for the most part, with the French in Morocco, the British in Egypt, and Sudan, the British and French in Libya after the Second World War, a system of protectorates was established, that is to say states were nominally independent but external policy was controlled by the protecting power and internal government was also often heavily influenced. The motivation for these arrangements was not simply to protect European trading and investment interests, they also included considerable strategic elements, as with British anxieties over the Suez canal route to India.

In the Middle East a not entirely dissimilar pattern of domination by outside powers, particularly by the British and the French, followed the dismantling of the Turkish Ottoman Empire at the end of the First World War. Some of the states that emerged here had ancient historical roots, although that was no guarantee, as in Iraq, of homogeneity and stability. Others were more reminiscent of the colonial states created in Africa south of the Sahara, in that the demarcation of Iraq, Saudi Arabia and Kuwait did not have exact and unambiguous territorial identity. At the beginning of the period with which we are concerned, nearly all the states of North Africa and the Middle East were truly independent, only a few minor areas such as the Trucial States had yet to achieve independence (as the United Arab Emirates, from Britain, in 1971).

The same kinds of questioning about the future that occurred for Black Africa also applied for North Africa and the Middle East. The oil wealth of Algeria, Libya and the Middle East was already known, as also the phosphate wealth of Morocco. It soon became apparent, however, that resource endowment, especially in water, was poor – particularly in Egypt in relation to its population, and in some other smaller states such as Jordan. The concern from the beginning for Western observers was for the security of Western interests, and particularly in the oil exporters. Given the proximity of the Gulf to the USSR and the fierce strategic struggle on the world's chess board between the USA and the USSR, it was inevitable that both superpowers should get heavily involved with regional states, hoping to establish patron–client relationships, and exclude each other.

Such a state of affairs alone would have been enough to guarantee instability and strife, but to it was added another overwhelming aggravation, the state of Israel, carved partly by force as a homeland for the Jews in 1948 out of the former British protectorate of Palestine. This short and violent war caused the original efflux of homeless Arab Palestinian refugees, stateless to this day. Israel's continued existence has been tested in several wars, each of which has dragged in outside powers and which has wrong-footed virtually all participants in some way or other. Israel's continued existence ought to be of no particular strategic concern to the outside powers, it has little of value to offer to the interests of the United States or Europe, other than its extremely remote utility as a last-ditch forward military base if at some future time no Arab state would countenance any connection whatsoever with the West. The Western involvement with Israel rests on other motives of self-interest and sentiment: that of the USA is based almost entirely on the large and powerful Jewish domestic lobby within the USA, and it is that which guarantees the continued external support.

Debates about the development of the Middle East have thus mostly been about the establishment of regional security pacts which simultaneously satisfy external interests, and about the elusive search for peace. Concern about the political viability of the states was never expressed in quite the same way as for Africa south of the Sahara. They were not on the whole seen as a novel product only of the last 50 to 100 years of European involvement (the exception to this may be the Arabian peninsula). Nor was their future economic development seen with the same anxiety, nor the same sense of responsibility. Most states appeared to have the resources or infrastructure as the basis for development, and most had elite groups who could provide some kind of leadership, however politically unlegitimized. In retrospect, perhaps a more sympathetic interest in the development of political institutions and in the developmental welfare of the people *per se*, would have served both the region and the developed world much better.

To the extent to which it is possible to say what has happened in the last twenty-five years, the rest of this book is dedicated. We do so by using a regional breakdown, grouping the states within the major divisions of Africa and the Middle East. There is one slight exception to this general arrangement – in that one state, Israel, is grouped non-contiguously with Turkey and Iran, the non-Arab northern and eastern border states of the Middle East. Within each region our authors are concerned with the changing population structures and issues of health, the changing patterns of urbanization and communications, the changing economies and patterns of trade, and the changing political and cultural structures, and the impacts that all this is having on the physical environment. Lest anyone think that this is a 'whole story' in most chapters there are also strong strictures on the limits of our knowledge.

REFERENCES

Andrae, G. and Beckman, B. (1985) *The Wheat Trap*, London and Uppsala: Zed Books.

Arden-Clarke, Sir Charles (1961) 'The West and Africa's challenge', *African Affairs* 60(241): 501–7.

Berg, E.J. (1965) 'The development of a labour force in sub-Saharan Africa', *Economic Development and Cultural Change* 13(4): 394–412.

Davis Fogg, C. (1965) 'Economic and social factors affecting the development of smallholder agriculture in Eastern Nigeria', *Economic Development and Cultural Change* 13(3): 278–92.

Dumont, R. (1962; 1988) *False Start in Africa*, London: Editions du Seuil.

Fieldhouse, D.K. (1986) *Black Africa 1945–80: Economic Decolonization and Arrested Development*, London: Allen & Unwin.

Fordham, P. (1974) *The Geography of African Affairs*, Harmondsworth: Penguin.

Frankel, S.H. (1960) 'Economic aspects of political independence in Africa', *International Affairs* 36(4): 440–6.

Hirschmann, A.O. (1965) 'Obstacles to development: a classification and a quasi vanishing act', *Economic Development and Cultural Change* 13(4): 385–93.

Minogue, M. and Molloy, J. (eds) (1974) *African Aims and Attitudes*, Cambridge: Cambridge University Press.

Perham, M. (1961a) 'The colonial reckoning', *The Listener* 66(1703): 795–8.

Perham, M. (1961b) 'African nationalism', *The Listener* 66(1704): 851–5.

Perham, M. (1961c) 'The politics of emancipation', *The Listener* 66(1705): 898–901.

Rennie, Sir Gilbert (1961) 'Prospects of division in Africa', *African Affairs* 59(239): 177–88.

Robertson, Sir James (1961) 'Sovereign Nigeria', *African Affairs* 59(239): 145–54.

Wallerstein, I. (1964) *The Road to Independence: Ghana and the Ivory Coast*, Paris: Mouton and Co.

2

THE CHANGING GEOGRAPHY OF SOUTHERN AFRICA

Deborah Potts

INTRODUCTION

Southern Africa comprises ten countries: Angola, Botswana, Lesotho, Malawi, Mozambique, Namibia, South Africa, Swaziland, Zambia and Zimbabwe (see Figure 2.1). The reader will avoid some potential confusion in what follows by noting the distinction between the region, known as southern Africa, and one of its constituent countries, South Africa, or the Republic of South Africa. All the countries of the region have experienced significant changes in their political, economic and social geography in the past thirty years. Many changes are similar to those which occurred elsewhere south of the Sahara in the same period, but one theme distinguishes this region from the rest of the continent. This is the extensive influence of South Africa on all the countries in the region. Southern Africa has experienced massive political, economic and military destabilization since the mid-1970s which, in different countries, has been largely or partly inspired by South Africa. In addition South Africa has strong economic links with the rest of southern Africa in the form of trade, transport and migrant labour.

South Africa's political system of apartheid has led to international censure and increasingly severe economic sanctions. It is partly the defence of this system which led to South African aggression in the region. All the governments of the region, with the partial exception of Malawi and the current government of Lesotho, have been strongly opposed to the apartheid government in South Africa. Thus one of the aims of the regional organization SADCC (Southern African Development Co-ordination Conference) is to lessen ties with South Africa. However, a number of factors,

Figure 2.1 Southern Africa

including geographical realities and the impact of destabilization, have made this process very difficult. The significance of South Africa's influence on the region's changing geography will be a recurrent theme in this chapter.

POLITICAL CHANGE SINCE 1964: IDEOLOGIES AND EXTERNAL AFFILIATION

Profound changes have been wrought in southern Africa's political geography in the past three decades. In 1960 none of the countries was independent, with the exception of South Africa which had been independent since 1910. Portugal still ruled Mozambique and Angola, whilst Britain ruled the other countries with the exception

13

of Namibia, which was governed by South Africa. Malawi and Zambia gained their independence in 1964; Botswana and Lesotho followed in 1966, and Swaziland in 1968.[1] Shortly after independence both Malawi and Zambia became one-party states, under the leadership of Hastings Banda and Kenneth Kaunda respectively. Remarkably, Malawi still remains under the control of its first leader, although the move towards multi-partyism in Africa finally led to Kaunda's replacement by Chiluba at the end of 1991. These countries' political development both bear the stamp of their leaders' ideology: Banda is conservative, authoritarian and essentially capitalist in orientation, whilst Kaunda professed his own ideology of humanism, which in theory combines elements of both socialism and capitalism. Both countries' economies have been strongly influenced by government intervention, and Zambia also nationalized the copper mines which dominate its economy. Both Swaziland and Lesotho were monarchies at independence. The new king, crowned in 1986, still plays an important role in Swaziland, where ideology has remained conservative and pro-Western, although there is some conflict between traditional and more liberal political factions. Lesotho is presently ruled by a military government, and the king has little role. Botswana is often held up as a shining example of a democratic, multi-party state in Africa. The electorate has persisted in returning the Botswana Democratic Party at every election. The government is again broadly capitalist and pro-Western in orientation, although it has combined this with a very definite, if carefully managed, alignment with its northern neighbours against South Africa's policies of apartheid and regional aggression.

The fight for independence and its ideological legacy

In Zimbabwe,[2] Angola, Mozambique and Namibia, where there were significant white minorities, independence was won only after long and bloody wars of liberation. Institutionalized racial inequality was characteristic of the whole region before the 1960s, but in these countries where the white minority had tried to maintain its privileged position by force, in the face of the sweeping tide of African independence and world opinion, black anger about the iniquities of such inequality was greatly heightened. In addition, the opposition expressed by the Western capitalist world to Portugal's and South Africa's actions in Angola, Mozambique and Namibia, was often largely rhetorical. The most powerful liberation move-

14

ments which eventually formed the independent governments in the four territories received direct support for their struggle only from communist countries, including the Soviet Union, China, East Germany and Cuba. This support naturally influenced the ideological development of the liberation movements. Furthermore, mistrust and antipathy towards the West was increased by the actions of the United States and Great Britain in particular, in watering down or vetoing innumerable calls in the United Nations and other bodies for sanctions against South Africa and Portugal. Such factors help to explain the adoption of radical left-wing political philosophies by the liberation movements. When Angola and Mozambique finally gained their independence in 1975, the Popular Movement for the Liberation of Angola (MPLA) and the Front for the Liberation of Mozambique (FRELIMO),[3] the parties which took over from the Portuguese, were both Marxist-Leninist and many of their policies stem from this ideology. There was widespread nationalization in both countries and their external affiliations were strongly orientated to the Eastern Bloc. Both countries became members of the Front Line States (FLS), an informal grouping of countries including Zambia, Botswana and Tanzania, which have played a major diplomatic and lobbying role in trying to end white minority rule in the region. South African aggression against the FLS, which includes support of anti-government forces in Mozambique and Angola, has been a major factor influencing their political and economic development.

The independence of Zimbabwe

International action against Rhodesia was more definite than against the other white-ruled states, since the white settler minority had defied the British by illegally proclaiming in 1965 a Unilateral Declaration of Independence (UDI). It was this flouting of the colonial power, rather than the continued exclusion of most Rhodesian Africans from political power, that led to the rapid imposition of comprehensive and mandatory international economic sanctions against Rhodesia. Although the impact of sanctions was lessened by South Africa's and Portugal's help in evasion, by the late 1970s they were beginning to take their toll of the economy and white morale (Minter and Schmidt 1988). After 1975 Mozambique not only supported sanctions but also assisted the liberation struggle by giving territorial bases to the Zimbabwe

African National Union (ZANU); Zambia similarly aided the Zimbabwe African People's Union (ZAPU). Their assistance and attempts to apply sanctions resulted in significant costs to their own political and economic development; the impact on Zambia's external trade routes was particularly severe.

As it became increasingly evident that white Rhodesia could not survive, and as government attitudes in the West changed, it came under pressure to compromise. It was also recognized that the longer the war was allowed to drag on, the more radical the eventual black government was likely to be. This latter factor also encouraged South Africa to push white Rhodesia towards a solution. Britain convened a conference at Lancaster House in London in 1979 at which all the major political parties were represented. A new constitution was drawn up, and arrangements made for new elections. At the time ZANU and ZAPU formed a coalition, the Patriotic Front (PF), and the constitution they accepted represented a major compromise of many of the radical principles which elements of both parties espoused. Capitalist property relationships were protected, and complex regulations prevented rapid land reform.[4]

Despite the hopes and machinations of Western nations and South Africa (Frederikse 1982; Stoneman and Cliffe 1989: 34), ZANU, perceived as the most radical party, won the 1980 elections

Table 2.1 Changes in the white population of selected SADCC countries

Country	Year	Population ('000s)
Angola	1960	173
	1974	330
	1976	30
Mozambique	1955	55
	1974	250
	1976	15
Zimbabwe	1962	220
	1982	148
Malawi	1966	10
	1977	6
Zambia	1961	75
	1969	43

Sources: Census data, and estimates for 1970s population in Angola and Mozambique.

with a large majority, and Mugabe, a professed Marxist, became Prime Minister. However, two major factors combined to ensure that Zimbabwe's approach to implementing socialist ideals has been extremely cautious. First, the conditions of the Lancaster House constitution tied the government's hands in many ways – they could not just be ignored for fear of Western financial sanctions. Second, President Samora Machel of Mozambique urged Mugabe to adopt a reconciliatory posture towards the white population. In both Mozambique and Angola most of the whites fled at independence (see Table 2.1), taking with them significant assets and frequently sabotaging what they had to leave behind. Since African education had been grossly neglected, nearly all skilled labour was also lost. The economic impact of the exodus had thus been devastating.

However, Zimbabwe's positive record on reconciliation between whites and blacks also reflected a genuine desire to create a successful multi-racial society. Much has been done to redress institutionalized aspects of racism in the country: all facilities are now open to everyone, legal restrictions on black urbanization have been abolished and laws on residential segregation repealed. In addition greater resources have been allocated to the African farming sector and to African education. These changes have not altered the fundamental nature of the political economy however, which remains capitalist, with a strong private sector much of which is foreign-controlled.

ZAPU and ZANU signed a unity accord in 1988, ending strife between government forces and ZAPU dissidents which had developed in the southern provinces after independence. Until 1991 government statements still referred to a commitment to eventual socialist transformation, and Mugabe still publicly supported the creation of a one-party state. However, neither objective is likely to be realized: many key politicians are opposed to one-partyism, and there is now an emergent black bourgeoisie (including many government members in defiance of ZANU's leadership code on private capital accumulation) with a vested interest in a capitalist economy.

The development of South African regional policy

South Africa's foreign policy actions towards its neighbours have had a decisive influence on regional political developments since the mid-1970s. The establishment of Marxist governments in the former Portuguese colonies marked a turning point. Both white-ruled

South Africa and Rhodesia felt threatened by these events. Until then South Africa had perceived Angola and Mozambique, along with Rhodesia and Namibia, as buffer states against the spread of black majority rule from the north. The political ideology espoused by the new nations was abhorrent to apartheid South Africa, which together with Rhodesia feared, correctly, that their own liberation movements would gain important new, and uncomfortably close, sources of pyschological and physical support. The white-ruled states reacted aggressively to the threat. Rhodesia bombed ZANU bases and refugee camps in Mozambique, and also ZAPU camps and bases in Zambia. It also established and supported an anti-Mozambican government force which included dissident FRELIMO ex-combatants, aided by right-wing Portuguese elements. This force, known as the Mozambican National Resistance (MNR),[5] attempted to destabilize Mozambique by destroying infrastructure such as railways, pipelines and electricity lines, and attacking villages, targetting government employees such as teachers and health workers in particular.

South Africa's interventions were even more direct. As it became clearer that the MPLA was going to come to power in Angola, the SADF (South African Defence Forces) invaded from Namibia in October, 1975. They were encouraged in this action by the foreign policy stance of the USA, which perceived that events in southern Africa were a direct threat to American interests. The USA had major private oil interests in Angola, but of greater concern was the perception of the risks of a possible shift in the region towards the Soviet bloc. A major consideration was the global significance of southern Africa's resources: the region contains large reserves of many important strategic minerals, including the platinum group of metals vital to high-technology defence and space developments, and South Africa has the world's largest gold reserves. The major alternative source of some of these minerals was the Soviet Union. In addition, significant proportions of the West's oil from the Middle East was shipped around the Cape. Western nations, particularly Britain, also had major financial and capital investments in South Africa. South Africa's invasion of Angola occurred therefore in a climate which led it to expect US support. The USA was directly supporting Jonas Savimbi's UNITA (National Union for the Total Independence of Angola),[6] a rival movement to the MPLA, which was seen as a softer option than the Soviet-backed MPLA. The SADF succeeded in moving northwards through two-thirds of Angolan territory, but American support was not forth-coming as Congress refused to allow direct US involvement on foreign soil, after the Vietnamese débâcle. The South Africans, in the face of Cuban military support for the MPLA, withdrew.

South Africa's regional foreign policy, adopted in 1978, was known as the total strategy and was couched in terms of defence against the 'total communist onslaught' . This policy involved the use of every means available to prevent the successful establishment and operation of the South African and Namibian liberation movements in neighbouring countries. The means included the direct military and financial support and organization of surrogate anti-government forces in the surrounding countries, in order to destabilize their governments.[7]

UNITA and the MNR have been the main tools of South African-backed destabilization. Armed anti-government factions were also supported in Zimbabwe in the early 1980s (Super-ZAPU), and in Lesotho from the end of the 1970s (the Lesotho Liberation Army). A major aim was to bully South Africa's neighbours into denying bases to the ANC and other organizations. It was also in white, capitalist South Africa's interests to prevent successful economic and political developments in socialist Angola, Mozambique and Zimbabwe. The reasons are twofold: first, such success would be a further boost to the exiled South African liberation movements which all espoused some form of socialism. Second, South Africa had a strong vested interest in preventing progress in these countries, because it could then hold their failures up as 'proof' to the Western capitalist world that black majority rule was disastrous. The purpose of such propaganda was to persuade Western governments, which were under domestic and international pressure to take effective action against South Africa over apartheid, that a white minority regime in South Africa was preferable to a post-apartheid, and probably socialist, black majority government. Destabilization of the region's economies and infrastructure, particularly the railways, also had the effect of pushing their northern neighbours into a much more significant degree of economic dependence on South Africa's industries and railways than would otherwise have occurred. The advantages were that South Africa could, to discourage the West from imposing sanctions, threaten counter sanctions aginst these countries, and it could manipulate its various trade and transport links to threaten and cajole its neighbours into not being too overtly hostile, as well as to destabilize their economies. During the 1960s and 1970s the government had nurtured hopes of creating a regional economic grouping CONSAS (the Constellation of States), in which political control of the surrounding countries would have been facilitated by the development of formal links with the Republic's dominant economy (see for example Geldenhuys 1981; Shaw 1981). The political orientation of independent Angola, Mozambique and Zimbabwe, and the creation of SADCC, however, put an end to this strategy.

The political impact of South African destabilization

There is no doubt that South Africa's destabilization policies achieved many of their objectives. In political terms they ensured that the polities of Angola and Mozambique have been rendered extremely weak, and often ineffective. Huge resources have had to be devoted to defence, and many development programmes have been difficult or impossible to implement. The extent of UNITA's and the MNR's dependence on South African support for their existence and success has varied and is a subject of debate (see for example Hall 1990; Smith 1990), but evidence suggests that the MNR probably would have faded away when Rhodesian support ended in 1980, if the South Africans had not stepped in (Hanlon 1986: 140; Smith 1990: 36). Its activities drove Machel to sign the Nkomati accord with South Africa in 1984, thereby tacitly acknowledging the Republic's regional dominance. He expelled the ANC but South Africa's support for the MNR remained, and the civil war has escalated. There is still at least covert South African support for the MNR which is now an extraordinarily vicious movement, lacking a central identity or political agenda. It is also aided by a number of foreign right-wing groups. Dissatisfaction with FRELIMO policies, particularly the undermining of traditional political and religious authority, and villagization, has led to occasional local support, but the MNR's tactics of abductions, rapes, massacres and wholesale theft and destruction have denied them any possible identity as a 'popular' resistance movement.

UNITA, on the other hand, does have a strong ethnic support base amongst the Ovimbundu of the central highlands and has a clearer political agenda, which broadly involves a return to a free market economy. Nevertheless without direct SADF intervention on a number of occasions, the civil war might not have continued, and some political compromise might have permitted the development of a more effective government.

Pressure from the international community on South Africa to end its illegal occupation of Namibia eventually led South Africa to link this issue to the ending of Cuban support for the Angolan government, a move resisted by the FLS. By the end of the 1980s the expense of the external 'total strategy', the upsurge of internal political resistance in South Africa and the military defeat of South African forces at Cuito Cuanavale in Angola in 1988 led to a change. It appears that the SADF was instrumental in persuading the South African government that the scale of its regional military interventions could no longer be maintained. This eventually led to Angola accepting the 'linkage' deal, the departure of the Cuban forces, and, on 1 April 1990, Namibia finally achieved its independence. It was Africa's last colony.

The governments of Angola and Mozambique have been engaged in discussions with UNITA and the MNR since 1989. In Mozambique these discussions have so far achieved very little and the MNR continues to operate to devastating effect. In May 1991 it seemed likely that a rapprochement was finally being worked out in Angola between UNITA and the government in Luanda.

South Africa was also involved in toppling Chief Jonathan of Lesotho in 1986, who was becoming increasingly outspoken about the apartheid regime, and replacing him with General Lekhanya and an essentially conservative military regime more amenable to South African influence. Zimbabwe has retained its strong stance against apartheid and South African regional destabilization. However, after ten years of threats and periodic sabotage, it also retains a strong sense of insecurity *vis-à-vis* its southern neighbour. Internal political security and popular support also suffer some damage from MNR activities in its eastern districts. The cost of defending the Beira corridor – Zimbabwe's rail, road and pipeline transit route through Mozambique without which its room for manoeuvre against South Africa would be even more limited – is also a political bone of contention within the country.

Changing ideologies or political pragmatism?

In the last few years political patterns in southern Africa experienced some important shifts in emphasis. By 1990 both Angola and Mozambique had adopted economic policy changes which introduced some free market forces into their previously centrally planned economies. Mozambique also officially dropped its adherence to Marxist-Leninist principles in the party congress of 1989. This shift undoubtedly reflects disillusionment with economic failures attributable to the over-bureaucratic and centralized approach typical of command economies. However, the degree and pace of change, particularly in Mozambique, also reflects pressure exerted by the West and its agents such as the IMF and World Bank. Aid donors have also played a part. This process of economic liberalization, partly induced by external forces, is currently widespread in Africa. At present, in countries which were so economically weak that there was little choice but to bow to the IMF's conditional programmes, it is difficult to ascertain to what degree these changes are indicative of a real ideological shift in government attitudes. IMF programmes have so far been implemented in Malawi, Zambia, Mozambique and Lesotho.[8] In 1987 Angola adopted its own programme,[9] but the pace and effectiveness of the changes have been rather limited so far. Zimbabwe is also implementing liberalization under its own control; its approach is typically cautious and does not really reflect any sudden ideological shift.

The other important changes in the region involve South Africa and Namibia. Although it is too early to characterize the ideology of the Namibian state, Nujoma's SWAPO party has publicly stated its continued commitment to some form of socialism. SWAPO does not have an overall majority, and it is probable that Namibian policies will display similarities to those in Zimbabwe, with a mixed economy and some redress of glaring racial inequalities. As in Zimbabwe the economy is presently dependent on white skills and foreign investment, and pragmatic considerations are bound to temper policy decisions.

In South Africa the election of President de Klerk in 1989 has ushered in sweeping political changes, which hopefully herald the end of the apartheid era. The liberation movements have been unbanned,[10] and the major black political leaders freed. There is now a political dialogue between the major political actors, without which real political change could not occur. De Klerk undoubtedly appears committed to change, repealing fundamental apartheid laws such as the Group Areas Act, the Native Land Act, and the Population Registration Act in February 1991. Concerns about the partisan behaviour of the security forces

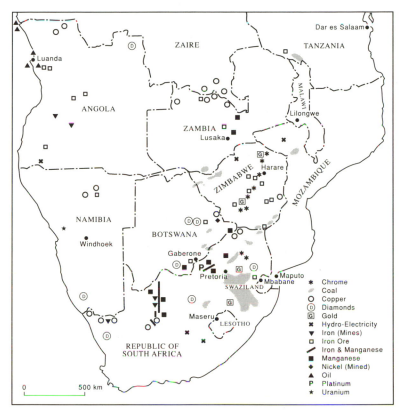

Figure 2.2 Southern Africa: mineral and energy resources

remain one of the important stumbling blocks to a long-term solution.

ECONOMIC CHANGES: AN OVERVIEW

Economic patterns at the beginning of the 1960s were partly the result of two unalterable factors: the known physical resource base (see Figure 2.2) and the individual countries' colonial economic history. These factors are still very influential and are also inter-connected, since the location of mineral reserves was an important determinant of the pattern of white settlement. This, in turn, was a decisive factor influencing the economic and political policies which shaped the different countries' economic geography. Two

23

Table 2.2 The racial division of land in South Africa, Namibia and Zimbabwe

% of total	South Africa 1960	Namibia 1960	Zimbabwe 1960	Zimbabwe 1990[a]
'white' land	87[b]	40	47	35
'African' land	13	50	47	65
white population	15	10	7	1
African population	75	90	93	99

Notes: percentages of land may not add to 100 per cent because of 'other' categories not shown (e.g. forest reserves, game reserves).
[a] The 1990 situation is shown only for Zimbabwe, as there has been little change in the other two countries. In this year 'white' land includes commercial farming land owned by other races. African land includes resettlement areas.
[b] This 87 per cent includes a small amount of land owned by Asians and Coloureds, who make up the remaining 10 per cent of the population.

Table 2.3 Population by race and ethnic group in South Africa 1980

Race/ethnic group	number ('000s)
BLACK	21,079
of which[a]	
Ndebele	545
North Sotho	2,350
South Sotho	1,938
Swazi	737
Tsonga	985
Tswana	2,083
Venda	522
Xhosa	5,229
Zulu	5,495
ASIAN	818
COLOURED	2,687
WHITE	4,526
of which	
Afrikaners[b]	3,000 (estd)

Sources: Smith (1985) Table 4.1; Steenkamp (1989).
Notes: [a] These ethnic categories are those used by the South African government, and are the only categories for which data are available. However, it should be noted that this is a very crude classification.
[b] This is a rough estimate only of those for whom Afrikaans is their first language.

fundamental aspects of economic patterns in South Africa, Rhodesia and Namibia were shaped by policies designed to create and maintain racial inequality. First, the land was divided between whites and blacks in a completely unequal manner (Tables 2.2 and 2.3), and white areas received the bulk of transport and other infrastructural investment. Furthermore, land *ownership* in both rural and urban areas was almost entirely a white preserve, and rural land designated for African occupation was mainly held under ⌐onditions of traditional, communal tenure. Second, a plethora of restrictions and impediments was implemented to limit the occurrence of permanent African urbanization, and encourage the circular migration of men without their families. *All* the major urban areas were designated as 'white'. These policies naturally had an immense impact on, *inter alia*, population distribution, patterns of agricultural development, transport and infrastructural geography, rural–urban migration, and urban morphology. As yet little has happened to alter these patterns in South Africa and Namibia, and the experience of Zimbabwe, ten years after independence, indicates that it is very difficult to transform the economic geography inherited from a racist state. A major problem is the prohibitive expense, particularly when land expropriation without compensation is, as in the case of Zimbabwe, deemed politically infeasible.

Changing economic patterns

The change in economic patterns which has occurred in the past thirty years has tended to be greater in those states which had to fight for their independence. The following analysis of change will look at agriculture, the industrial and mining sector, and urbanization and migration in turn. The impact of destabilization, and the changes in transport geography which have largely stemmed from it, merit separate analysis. South Africa's economic geography is unique and also requires separate consideration. Table 2.4 illustrates the changes in the sectoral distribution of the different countries' GDP.

Agriculture

It is difficult to generalize about the changes in agriculture as the resource base and the policies implemented over the past thirty years vary. In some countries there has been continuity from the

Table 2.4 GDP by sector (%) in southern Africa 1965 and 1987

Country	Primary Agriculture '65	'87	Mining '65	'87	Secondary Industry[a] '65	'87	Manufacturing '65	'87	Tertiary Other '65	'87
Angola[b]	13	11	10	18	27	—	8	11	60	—
Botswana	34	3	0.5	44	19	57	12	6	47	40
Lesotho	65	21	1.6	neg	5	28	0.7	15	30	51
Malawi	50	37	0	0	13	18	10	15[c]	37	45
Mozambique[d]	20	50	7	—	—	12	13	—	—	38
Namibia[e]	—	13	—	28	—	44	—	7	—	41
South Africa[f]	10	5	13	—	42	45	22	23	48	50
Swaziland	35	24	12	—	33	30	9	20	32	46
Zambia	14	12	41	15	54	36	6	23	32	52
Zimbabwe	18	11	7	8	35	43	20	31	47	46

Sources: World Bank (1989); Peet (1984); Fransman (1982); *Financial Times* (1990).
Notes: [a] Industry includes mining, manufacturing, transport and communications, utilities.
[b] Angola: data are for 1971 and 1980 rather than 1965 and 1987, and are taken from Peet (1984) since World Bank data for Angola are inaccurate.
[c] 1981 data.
[d] Data are for 1970 and 1987.
[e] Data are for 1988.
[f] Data are for 1965 and 1985.

colonial past, although where there have been problems of land shortage these have tended to be exacerbated by population increase. In others there has been fundamental change. Broadly speaking the former are those countries which attained independence through the 'normal' process in the 1960s, and the latter includes those where independence was delayed, and more radical governments emerged from the struggle against white minority rule.

Amongst the former group, Malawi and Swaziland have had more successes in their experience of agricultural development than Zambia, Lesotho and Botswana. As shown in Table 2.5 export crop production has tended to increase in these two countries. Malawi has also taken great pride in maintaining its self-sufficiency in food. None of the other countries has achieved this, although Zambia is on occasions self-sufficient in maize. Maize, the staple food of the entire region, is by far the most important crop grown. In each country most peasant households try to grow at least enough maize for subsistence. However, many are unable to achieve this for a variety of reasons. Physical factors include land shortage in Malawi and Lesotho, and low and unreliable rainfall in Botswana. Government pricing policies and the parastatals (government-owned

Table 2.5 Agricultural production in southern Africa 1965–87 (thousands of metric tonnes)

Country	Date	maize production	cereal imports[a]	EXPORTS tobacco	sugar	cotton	tea	coffee
Angola	1965	450	116	2	—	5	—	159
	1987	300	280	0	—	1	—	11
Botswana	1965	2	79	—	—	2	—	—
	1987	2	137	—	—	0	—	—
Lesotho	1965	110	83	—	—	—	—	—
	1987	90	94	—	—	—	—	—
Malawi	1965	890	116	17	0	—	13	—
	1987	1228	11	61	66[b]	—	33	—
Mozambique	1965	390	128	—	95	31	—	—
	1987	300	406	—	25	20	—	—
Swaziland	1965	32	44	—	100	1	—	—
	1987	92	44	—	436	5	—	—
Zambia	1965	800	169	10	2	—	—	—
	1987	954	150	6	24	—	—	—
Zimbabwe	1965	822	75	123	254	1	—	—
	1987	1094	71	100	283	66	—	—

Sources: World Bank (1989); Lipton (1988).
Notes: Main agricultural exports only are shown.
[a] Cereal import data are for the three years 1970, 1980 and 1986.
[b] Sugar production continued to rise in Malawi from 158,000 tonnes in 1980, to 184,000 in 1987. The fall in exports was largely caused by transport problems due to destabilization.

corporations and boards) involved in marketing agricultural produce and providing inputs, have had a major, and often debilitating impact on the agricultural sector. Prices paid to the farmer for domestic and export cash crops have often been much lower than those obtained when the parastatal sells them on. This process may be regarded as an unjust transfer of revenue from farmer to government, but governments have to derive revenue from somewhere, and similar policies were used during the colonial period. However, the extent to which, effectively, the sector is paying an indirect tax has increased substantially. In addition, the funds allocated to peasant agricultural development have not reflected its importance, either in economic or welfare terms. Furthermore, parastatals have often been inefficient or even corrupt. If they mismanage their vital role in buying and transporting crops, and providing timely inputs, this may be an even greater disincentive to production than prices. These problems, by no means unique to southern Africa, can be illustrated by a few regional examples.

In Malawi, ADMARC (the Agricultural Development and Marketing Corporation) is the parastatal which buys and markets peasant crops, and provides inputs. Maize prices paid in relation to their real local value have fluctuated since independence. Despite some substantial price rises in the 1980s, the advantage for farmers has often been offset by inflation and large fertilizer price rises. Other peasant cash crops like cotton, groundnuts and tobacco have tended to fare much worse in terms of official prices, however. Continuous government exhortations to 'grow more maize' and 'work hard in the fields' have had some effect in generating maize surpluses: undoubtedly the President and his party's agents have not made the mistake of ignoring farmers, at the rhetorical level at least. The President even makes annual tours of all the districts to examine the maize harvest, and he makes lengthy public pronouncements about what he finds. ADMARC has been relatively efficient in maintaining a nationwide network of marketing depots, and crop prices were not regionally differentiated. This amounts to subsidizing more remote producers, and helping regional equality. On the negative side, the allocation of government resources for agricultural development has favoured large private estate development. In some localities, aid-assisted projects aimed mainly at peasant production of export crops have also been developed, but these have helped only a minority (Smith 1990). Many of the new private estates grow tobacco and are owned by the Malawian elite, whose entry into this sector has been subsidized by the government. Some of their land was previously in the hard-pressed peasant sector. The estates employ large numbers of labourers on a seasonal basis and the practice of paying very low wages is backed by government policy. A new development since independence is massive rural–rural migration as many peasant farmers, desperate for cash, now spend part of their year on such estates (Kydd and Christiansen 1982; Christiansen 1984).

Despite its good record on national food production, malnutrition and poor health are still major problems in Malawi, which has a very high infant mortality rate.[11] Poverty often drives farmers to sell some maize at harvest time, even when they do not have a surplus. Later, when they are forced to buy maize, prices have risen, and often they go short of food (Smith 1990). The country is suffering from a serious land shortage, given current agricultural technology. Diminishing plot sizes has led to overuse of the land and decreasing soil fertility. It is no longer possible to increase output by bringing more land under cultivation, so Malawi's record of self-sufficiency in food is in jeopardy. Intensification of production requires increased inputs, but generally speaking government policy towards the small-scale sector has paid insufficient attention to this problem. The adoption of high-yielding hybrid maize would be a partial solution, but seeds and vital inputs are far too expensive for most farmers unless the government allocates major new resources to help them implement the new technology. The proportion of land under high-yielding varieties (HYVs) actually fell during the mid-1980s, and was estimated at only 3–4 per cent of the cropped area in 1987 (Kydd 1989: 113).

The record of parastatal inefficiency has been worse in Zambia, and in general the agricultural sector has suffered from government neglect, despite the fact that Zambia has significant untapped agricultural potential: unlike some other countries in the region there is no national land shortage. Analyses of Zambia's agricultural development since independence recount a litany of overstaffed parastatals, poorly managed transport, lack of inputs and credit, and a demoralized and cynical peasantry. The inevitable result is that many small farmers, whilst maintaining subsistence production, are wary of investing in cash crops which rely on government marketing (see, for example, Pottier 1988). The agricultural sector has also been hit by Zambia's general economic decline which was partly externally induced by factors such as regional destabilization and a calamitous fall in world copper prices.

In Angola and Mozambique all sectors of agriculture have been in crisis since independence. Much of the large-scale sector came under government control as the white settlers left. The extreme shortage of managerial and technical expertise to run these farms and plantations, and of foreign exchange to maintain capital investment, led to major production losses. In the peasant sector governments tried to implement a number of policies which were viewed as progressive. These included encouragement of collective farming which was expected to lead to higher output and incomes, villagization (the planned concentration of a dispersed rural population in new rural centres) in order to facilitate provision of services and political education, and the replacement of traditional structures of village authority (e.g. tribal headmen) with elected committees. As in many other countries where similar policies have suddenly been imposed on traditional communities, these changes were often resisted. An even more significant factor was the general decline after Independence of both the Mozambiquan and Angolan economies. Wars in both countries have effectively destroyed the rural economy in many areas, and the crucial transport links between producers and markets. Millions of people have been displaced, and hundreds of thousands have died because of food shortages. Hunger induced by drought and war has been difficult to alleviate with food programmes in the war zones. There are around a million Mozambican refugees in surrounding countries, and Mozambique is reliant on food aid. The infant mortality rates in these countries may be the highest in the world, although data are scarce. Many fields are mined and tens of thousands of farmers have lost limbs or died whilst trying to cultivate. The real future of the agricultural sector in both countries now depends on peace and stability.

The migrant labour system and agriculture

Many peasant households throughout southern Africa are partly or almost wholly reliant on migrant wage remittances.[12] The migrant labour system (Low 1986; First 1983; Murray 1981) has been an important factor in peasant agricultural development in every country, with the partial exception of Angola, for a century or more. Oscillating migrant labour in which workers must return periodically to their rural 'homes' is an important characteristic of this system. Labour from South Africa, Namibia and Zimbabwe tends to remain within the home country, and in Zimbabwe the contract nature of this movement has largely ended. Contract migrant labour from the other countries tends to go to South Africa nowadays, although Zimbabwe was an important destination in the past. The absence of key household members can undermine the productivity of small farms. One problem is often the shortage of certain types of labour. The migration of labour has had a devastating impact on social and welfare conditions in those communities which are most heavily involved. In such communities the permanent population is dominated by women, and young and old dependants, and the dependence of rural households on migrant wages is very high. Today, this situation is particularly characteristic of the homelands of South Africa and Namibia, and Lesotho.

The major change that has occurred in the past thirty years has been South Africa's policy of reducing its foreign migrant labour intake and replacing it with local migrant labour from its homelands. This policy, which began in the 1970s, hit Malawi and southern Mozambique first, and from the end of the 1970s also led to reductions in recruitment from Botswana, Lesotho and Swaziland. There is little information on the impact of this on farming communities in southern Mozambique, which have any-way been affected by the civil war there, although the assumption is that it would have exacerbated rural poverty. In Malawi many workers appear to have been absorbed in part-time seasonal work on new tobacco estates at much lower wages (Christiansen and Kydd 1983). The situation has grave implications for Lesotho whose whole economy is dependent on migrant remittances. There, unemployment and internal rural–urban migration are rising, and many rural households are facing destitution.

The land issue in southern Africa

The most significant issue in the agricultural sectors of South Africa, Namibia and Zimbabwe, where land alienation by whites occurred on a massive scale, is that of land reform. The shortages of land which peasants face in these countries are artificially induced, and, unlike Malawi and Lesotho, for example, there is significant scope for easing the shortages by redistributing land from the white-owned commercial farming sector to the African peasant sector. The likely political and economic impact of such redistribution are matters of intense debate, and land reform is one of the major items on the political agenda in these countries. As yet there has been no change in the racially unequal distribution of land in South Africa and Namibia.

In Zimbabwe some redistribution has already occurred. An important cause of support for the war for liberation was peasant anger over land inequality, and the government has a political responsibility to its electorate to effect some land redistribution. Government ideology is also sympathetic to a more egalitarian pattern of land holding, although after ten years of cautious economic policies which have not materially altered the capital-ist orientation of the economy, and a concerted lobby by the commercial farming sector, it is hard to judge the present degree of commitment to land redistribution. In 1982 the government announced plans to resettle 162,000 peasant households, but the plans had to be cut back and by mid-1989 only 52,000 had been resettled. The terms of the Lancaster House agreement meant that resettlement could occur only on land which was offered for sale, and purchased, at market prices, remittable in foreign currency. The Lancaster House agreement had led the PF to expect significant foreign assistance with land purchase, but this was not forthcoming.

Financial constraints are by no means the only ones that limit changes in the pattern of land distribution in Zimbabwe. The white farmers are well organized politically, and the Commercial Farmers' Union argues strongly that changes in land-use from large-scale private farms to small-scale peasant farms with communal tenure would be economically disastrous. The assumptions behind such arguments are that this is likely to lead to significant falls in national agricultural output – on the grounds that peasant farmers are less efficient – and the replication of the land degradation which is characteristic of the communal areas.[13] These arguments are grossly exaggerated. First, there is evidence world-wide that peasant farmers can be highly efficient and productive; often their systems of land-use are more intensive than large-scale farmers. Second, the fundamental cause of land degradation in the communal areas is over-use of the land for grazing and cropping because there are too many people there; resettlement areas need not replicate these population densities. Finally, and most compellingly, the African small-scale agricultural sector in Zimbabwe since independence has achieved large increases in output of many crops, and is the most successful in the region. A 1988 British ODA report on the resettlement programme estimated that it had provided an economic return of approximately 21 per cent, and had 'proved a considerable success' (cited in Palmer 1990: 173). Data on Zimbabwe's agricultural sector are presented in Table 2.6. Another factor in the Zimbabwean land debate is the under-utilization of land by some white farmers, who are unable, or unwilling, to bring all their land under productive use. Definitions of 'productive use' vary: for example some land is used only for timber offtake from the indigenous woodland, and there is suspicion that, in some cases, game ranching has become a convenient bandwagon to ward off unwelcome claims that land is under-utilized.

Other strands to the land reform debate include, *inter alia*, the question of what is the most politically and economically desirable form of tenure in resettlement areas, who should be prioritized for resettlement, and whether present peasant practices regarding land-use need to be changed, and if so, how much and by what means.[14]

In 1990 President Mugabe stated that perhaps 50 per cent of remaining commercial farming areas would have to be purchased

Table 2.6 Crop sales by agricultural sectors in Zimbabwe 1970–90

	% marketed crop by sector					
	Maize		Cotton		Total crop sales	
Intake year	LSCS	SSS	LSCS	SSS	LSCS	SSS
1969–70	—	—	—	—	97	3
1974–75	96	4	—	—	94	6
1979–80	92	8	81	19	94	6
1984–85	59	41	55	45	79	21
1987–88	45	55	49	51	78	22
1988–89	40e	60e	<50e	>50e	81	19
1989–90	44	56	40	60	—	—

Sources: Zinyama 1986; FT Zimbabwe Surveys 1981 and 1989; Stoneman EIU
　　　reports 1989 and 1990; African Business Zimbabwe Survey May 1989; CSO
　　　Quarterly Digest of Statistics September 1989.
Notes: LSCS: Large-scale commercial sector (i.e. exclusively white at independence,
　　　and thereafter includes some black commercial farmers); SSS: Small-scale
　　　commercial and communal sectors; e – estimate; — no data.
　　　Values of sales of maize and cotton are for crop years (i.e. from harvest to
　　　harvest); value of total sales are for calendar year.

for resettlement, causing an outcry from the commercial agricultural sector. Whether this will occur remains to be seen, and it is
probable that the process of redistribution will remain a gradual
one.

Industrialization

The share of manufacturing and mining in GDP varies widely
between countries. In most the share of manufacturing has increased in the past thirty years, but there have been no fundamental
shifts in the nature of their economies which still remain essentially
reliant on trading primary products for manufactured goods. In
Malawi, Mozambique and Swaziland the export sectors are dominated by agricultural products. As is common in Africa, increases in
manufacturing in Malawi and Swaziland are partly related to
processing of such products. For example tea, tobacco and sugar
processing are important elements of Malawi's manufacturing sector (contributing 53 per cent of total manufacturing value added in
1980), as are timber and sugar processing and fruit-canning in
Swaziland. Swaziland's manufacturing sector has also benefited
from South African investment to avoid sanctions. Malawi achieved
impressive rates of growth in its industrial sector in the early

post-independence era, but this largely reflected the extremely small size of the initial industrial base. Many of the simplest types of import substitution were quickly developed, but this sector is constrained by the small size and poverty of the domestic market.

Manufacturing industry has remained fairly insignificant in the Lesotho economy. A major infrastructural development – the Highlands Water Project – is currently under construction, after many years of negotiation with South Africa and lending agencies. The main purpose of this project is to ease neighbouring South Africa's water shortage in its industrial heartland in the Transvaal.

In Zambia, Namibia, Angola and Botswana minerals dominate exports. In the first two this dominance was long-established. In Angola oil exploitation has increased significantly over the past thirty years. It already accounted for around 40 per cent of export value at independence, and by 1985 this had increased to 96 per cent. Diamonds and iron also contribute to the industrial sector, although their production has fluctuated because of the war. In Botswana the economy has experienced a remarkable period of growth since independence, due mainly to development of diamond mining, and to a lesser extent copper, nickel and coal mines. Secondary industrial developments have been negligible by comparison. Botswana now has large foreign exchange reserves, and a freely convertible currency.

Economic policies in Portugal's former colonies experienced an important shift in the 1960s. Foreign investment was encouraged, and much greater infrastructural and industrial development occurred. These developments were largely politically induced, since the Portuguese hoped that economic development might help to defuse the liberation struggle, and give Western investors a stake in supporting the colonial regimes. The policy met with some success in both countries, and by 1975 their industrial sectors were considerably stronger than those inherited by most African governments at independence. In addition to its important mining sector, Angola, for example, had vehicle assembly and chemical plants, whilst Mozambique had a steel plant and an important textile sector. The development of the Cahora Bassa hydro-electric plant on the Zambezi river was a very significant investment by a foreign consortium, and South Africa contracted to purchase most of the generated power.

After independence much of the industrial sectors of both countries was nationalized. The oil industry in Angola was a partial exception, and the state-owned SONANGOL operates in partnership with a number of foreign oil companies, the most important of which is the American company Gulf Oil. The scale of nationalization, which included many small firms, was more extensive than the governments envisaged, and was partly necessitated by their former owners' departure. Manufacturing production declined significantly in both countries. Factors involved included the loss of skilled personnel, outright sabotage by former owners, and lack of foreign exchange for inputs. Sometimes workers had to be paid with the products they produced. Strict price controls also led to many economic distortions, and a huge black market sector emerged in both countries. The economic reforms of the late 1980s include deregulating some prices, and allowing more private sector involvement in production. As yet these appear to have had little impact on increasing industrial production, and the legacy of the region's wars and financial shortages are still major problems.

Zimbabwe's industrial sector is by far the most sophisticated and successful in the region (excluding South Africa). The basis of the manufacturing sector was already developed by UDI. The 1950s witnessed an economic restructuring of some significance, with the development of a range of manufacturing plants. The white settler government made major investments in the development of a cotton-ginning industry at Kadoma, and an integrated iron and steel plant, using domestically produced iron and coal, at KweKwe. A side effect of UDI and sanctions was the imposition of conditions of autarky (economic self-reliance) upon the economy. The manufacturing sector enjoyed extremely high levels of protection from foreign competition for fifteen years, and foreign companies could not repatriate profits from their frozen accounts, and tended therefore to reinvest. The result was rapid expansion of manufacturing capacity aimed at the domestic market. The value of manufacturing grew at 5.4 per cent per year from 1965 to 1980 and accounted for 25 per cent of GDP at independence. Eventually sanctions took their toll of the ageing industrial plant and towards the end of the UDI period the industrial sector was in increasing difficulties. Nevertheless the manufacturing sector's base had been broadened. The initial post-independence period saw a short-lived import boom which allowed some vital capital goods' replacement. Throughout the rest of the 1980s strict import controls have been imposed to help the balance of payments. On the one hand, this has meant further protection for manufactures for the domestic market, and there has been further growth. On the other hand, as during UDI, this has also imposed constraints on the import of many inputs because of foreign exchange shortages, causing periodic bottlenecks and shortages. The Zimbabwean government has also encouraged export manufacturing industries, in order to earn foreign exchange, and this has met with some success. The government is now committed to gradual liberalization of the economy. It is hoped that careful management of this process, as opposed to the wholesale and sudden liberalization which IMF programmes have imposed on other African countries, will limit the damage that foreign competition may have on industry by allowing a period of adjustment. Zimbabwe also has the advantage of significant mineral resources, and gold, chrome and nickel exports have contributed up to 50 per cent of export value during the late 1980s.

SADCC and the economics of destabilization

In 1980 SADCC was formed as a regional grouping. It now includes all the states in southern Africa (excluding South Africa) and Tanzania. Unlike other regional groupings in Africa, the development of intra-regional trade has not been a major priority.[15] The development of self-reliance is more important, and although regional politics have determined that the main thrust of SADCC's political and economic activities have been directed at lessening dependence on the Republic, this also includes the aim of lessening external dependence generally. SADCC has achieved some significant successes in raising international finance for major infrastructural projects in the region, and in lobbying against the impact of South African regional policies. These two aspects of SADCC's activities are interlinked, since much of the investment has gone into rehabilitating infrastructure destroyed by South African forces, or its proxy anti-government movements. The cost of South African destabilization has been phenomenal. Current estimates for the period 1980–8 are US$62.45 billion for the region as a whole, equivalent to more than double SADCC's combined GDP in 1988 (Smith 1990). In Angola and Mozambique many of the improvements in primary health and education effected by the independent governments have come to nought, as the schools and health centres have been destroyed. The retail sector in rural areas has also been a target. Electricity lines and oil pipelines are regularly disrupted. The cost of defence forces has been a major drain on both countries, and on Zimbabwe. Any attempt to understand the changes in the region's economic geography must start with consideration of the spatial impact of destabilization. The effect on the region's transport network has been so profound that SADCC has prioritized this sector for rehabilitation and development. Each state has a sectoral responsibility within SADCC[16] and the responsibility for co-ordinating transport development has been allocated to Mozambique, the country worst affected by destabilization and transport disruption.

Changing patterns of transport

The colonial patterns of freight transport were influenced by two major factors: the location of significant mineral resources, and freight rates on the various routes to the ports. Zambia and Zaire's copper, cobalt and other mineral exports were vital to the profitability of the railways, as were Zimbabwe's mineral and agricultural exports, and trade from and to the Transvaal in South Africa. There were a number of routes available to the landlocked countries of Central Africa (see Figure 2.1). Zimbabwe's trade could use two routes through Mozambique to the ports of Beira or Maputo. Zambia and Zaire could use the Benguela railway line which connected Zambia to Zaire and thence through Angola to the port of Lobito. There were also alternative northern routes via Zaire, or Zaire and Tanzania, although the latter involved trans-shipment across Lake Tanganyika. Malawi had a transit route through Mozambique to Beira, and another railway to the excellent deep-water port of Nacala was developed after independence. With the exception of Malawi, all these railways were connected, via Zimbabwe and Botswana, to the South African railway system. In the late 1970s the UDI government constructed a direct link to South Africa, bypassing Botswana.

The landlocked countries of Botswana, Lesotho and Swaziland had a natural geographically determined dependence on the South African railway system, although Swaziland developed a rail link to Maputo after independence. However, prior to the period of instability engendered by the struggle against white minority rule, the transit route states competed to carry the trade of the northern landlocked countries, and the routes chosen were determined by the best deals available. For example, for a period, all Central African copper freight went through Beira as special rates were offered.

About half of the Transvaal's freight travelled through Maputo, the province's nearest port, because of an agreement with the Portuguese which also involved migrant labour recruitment rights in southern Mozambique. The South African railways carried some freight for Zimbabwe and Zambia, including imports from the Republic, but they were at a competitive disadvantage because of the extra distances involved (see Figure 2.1).

In the past thirty years these historical patterns have been greatly changed. From 1965 Zambia's options were limited due to its hostility to UDI, and the impact of the Angolan liberation struggle on the Benguela line. After 1975 the routes through Angola and Mozambique were effectively closed for much of the time, as the railways have been primary targets for the anti-government forces. For a long period Zambia and Zimbabwe were almost entirely reliant on South African transit routes. Their dependence rendered them extremely vulnerable to South African economic destabilization through freight hold-ups, which regularly occurred. Although Zambia constructed the TanZam (or Tazara) railway to Dar es Salaam, with Chinese aid, in the 1970s, technical and port capacity problems limited its use. Malawi became entirely reliant on road transport to connect it to transport networks in Zimbabwe, Zambia or Tanzania.

During the 1980s SADCC implemented a number of projects to rehabilitate the region's natural transit routes, which have attracted significant amounts of western aid and assistance. The Beira rail, road and pipeline 'corridor' has acted as a lifeline to the sea for Zimbabwe, and has been defended by up to 10,000 Zimbabwean troops. Malawi's railways, the Limpopo route from Zimbabwe to Maputo, and the Benguela line are all currently being rehabilitated, and the capacity of the various ports is being improved. In 1990 the Beira and Tanzam lines are the most effective alternatives to South Africa for the land-locked states. In 1988 it was estimated that 40 per cent of Zimbabwe's exports were going via Mozambique. Cargo handled at Beira is increasing steadily, totalling 2 million tonnes in 1987, compared to about 1.4 million tonnes in each of the previous three years. Nevertheless South Africa is likely to continue to play a major transport role for the region until peace is achieved in Angola and Mozambique.

URBANIZATION AND RURAL–URBAN MIGRATION

As elsewhere in Africa, the countries of southern Africa have experienced rapid urbanization, resulting in a significant redistribution of their populations from rural to urban areas. Whilst there are data available to indicate the rate of urban growth in the 1960s and 1970s, the situation in the 1990s is more difficult to assess. City sizes are often estimated on the basis of growth rates projected from the 1970s, a methodology which can lead to significant inaccuracies. Table 2.7 shows the increase in urbanization levels in southern Africa from 1965 to 1988. Urbanization levels in Malawi and Lesotho remain low, whilst South Africa's and Zambia's levels are high by African standards. Rural–urban migration is the main cause of rising urbanization *levels*, but as cities become larger, the contribution of natural increase to urban *growth* expands.

Table 2.7 Changing urbanization levels in southern Africa 1965–88 (% of total population in urban areas)

	1965	1975	1988
Angola	13	15	27[a]
Botswana	4	13	22
Lesotho	6	11	19
Malawi	5	7	14
Mozambique	5	7	24[a]
Namibia	28	30	55[a]
South Africa	47	47[b]	58[b]
Swaziland[c]	17	18	nd
Zambia	23	34	54
Zimbabwe	14	19	27

Sources: World Bank (1990); O'Connor (1983) except where otherwise indicated.
Notes: [a] Recent reliable urban data for Angola, Mozambique and Namibia are unavailable, so these are guesstimates at best.
[b] After 1970 the South African state excluded the so-called independent homelands from its population statistics. The estimate in the 1975 column is for 1980, and is calculated to include the independent homeland population. The 1988 World Bank estimate presumably excludes this section of the population.
[c] Data from 1962 census, and ILO report for 1977; no World Bank data available.

The causes of rural–urban migration are basically economic. Broadly speaking they reflect a shortage of cash-earning opportunities in the rural areas. Land availability and the profitability of cash crops are two major factors influencing rural incomes. The expansion of education which has occurred throughout much of the region since independence has also equipped many people with skills which may receive a higher return in urban employment. The expansion of formal wage-earning jobs in the urban areas has not kept pace with employment demand however, particularly during the past two decades. Many migrants, and urban-born people, support themselves by working in the informal sector. Drought and war, sometimes in combination, have also led to rural–urban migration of a less voluntary and selective nature. Instability in the rural areas led to an influx of people to the urban areas in Zimbabwe at the end of the 1970s, and has also created sprawling semi-urban settlements in northern Namibia. Angola's and Mozambique's cities have undoubtedly grown rapidly because of war refugees, but data are scarce. South Africa's policy of reducing its foreign migrant labour force has also increased internal rural–urban migration in Lesotho and Botswana in particular.

Many people who move to work and live in town may return to the rural areas eventually. Living costs in old age, illness or poverty caused by unemployment or falling incomes, may cause this return migration, and forced migration caused by instability may be reversed when peace is established. This occurred in Zimbabwe, for example, where many refugees returned to the rural areas after independence and may occur in Angola and Mozambique if the rural areas become more stable. Real income levels for the poorest people in the urban areas of many countries have been reduced significantly during the 1980s. The economic stagnation experienced throughout Africa in this decade affected urban employment and often led to deteriorating living conditions for the urban poor, in terms of housing, transport and urban services. Austerity programmes imposed by the IMF in Zambia, Malawi, Lesotho and Mozambique, or implemented independently in Angola and Zimbabwe, have probably hit the urban poor harder than any other section of the population. Many prices have risen dramatically, and wage increases have not kept pace on the whole. This situation has tended to decrease the average rural–urban wage gap, which in turn may have reduced rural–urban migration rates. In addition, return migration rates may have increased. This picture of slowing urbanization is hard to support with data at present, but numerous reports, anecdotes and logical deduction suggest that it is occurring. Available evidence indicates that Zambia's urban areas are the most affected.

The nature of rural–urban migration and the process of urbanization in the region is made complex by the continuation, in some areas, of short-term circular migration. This pattern was entrenched during the colonial period as permanent urban residence for Africans was often regarded as undesirable. Longer-term or permanent migration became more common in the 1960s, but in Zimbabwe, Namibia and South Africa a panoply of legal restrictions on permanent, family migration continued to distort the age and sex distribution of the African urban population. People of working age are over-represented in this population, as many dependants remained in the 'homelands' or 'tribal trust lands', and the sex ratio is severely imbalanced in favour of men. This pattern has now been redressed in Zimbabwe: urban growth is estimated to have been very rapid since independence, and many migrants now bring their families with them. Nevertheless, many migrant families who still have some rural land try to maintain production on it. A variety of strategies are used to achieve this: in many cases wives still spend the rainy season 'at home', and many others still remain in the rural areas on a permanent basis. Most migrants also plan to return to the rural areas in the long term (Potts and Mutambirwa 1990). But where land is in short supply, as in Zimbabwe, Malawi and Lesotho, the potential for becoming trapped in urban destitution is probably growing.

Changing urban and industrial patterns in South Africa

Economic activity in South Africa is concentrated in four major metropolitan areas: the Pretoria-Witwatersrand-Vereeniging triangle (PWV), Durban, Port Elizabeth and Cape Town. This basic pattern is long-established and can broadly be accounted for by the normal operation of economic and geographic forces as elaborated in the usual theories of industrial, urban and transport location. Mineral location, particularly the gold reef, and port development, in a country highly dependent on trade, combined with the historical patterns of white settler migration into the country's interior, have been important influences on the country's economic geography. The spatial expression of urban and industrial developments is the subject of a body of geographical literature (e.g. Fair 1982; Davies and Cook 1968; Browett and Fair 1974). Beyond this underlying pattern however, the country exhibits a number of features which in many cases run directly counter to theoretical expectations. The explanation for these features lies in the unequal division of land between Whites and Blacks,[17] a range of segregationist and urban influx control policies from the pre-apartheid era, and, since 1948, the implementation of various apartheid policies. In the apartheid era the former 'native reserves' which were the only areas where

Blacks had land rights, became homelands or Bantustans. African political and citizenship rights in white South Africa were removed, and shifted to these areas. All the homelands have self-governing status, although their revenues are highly dependent on grants from the government in Pretoria.[18] In the past twenty years the Transkei, Bophutatswana, Ciskei and Venda were also persuaded to take 'independence' but their status is unrecognized outside South Africa.

Mapping almost any modernization or welfare indices such as income, industrial value-added, wage employment, transport and communications, infant mortality rates, nutritional levels, or school enrolment rates would identify the homelands as spatial anomalies. If they are excluded, the country's spatial economy can be expressed and evaluated in terms of core-periphery theory (Browett and Fair 1974) and to some degree exhibits features found in many developed, capitalist countries. The fragmented homelands are generally the 'periphery of the periphery'. Unlike most 'development surfaces' where the contours express a gradually descending slope into the periphery, the borders of the homelands are usually marked by a sudden drop.

In the absence of apartheid, natural economic and geographic forces would undoubtedly have led to a marked shift in South Africa's population distribution, with rapid out-migration from the overcrowded and underdeveloped homelands to the white cities. However, in South Africa this expected pattern was reversed over this period until 1986, when some urban influx controls were dropped. The proportion of the Black population in the homelands increased from 38 per cent to 54 per cent between 1960 and 1980, as shown in Table 2.8. As a result the crisis of underdevelopment in the homelands has deepened. There are many factors involved in this shift in population which includes, inter alia, success in the implementation of influx control measures, the forced relocation of millions of blacks from white urban and rural areas, forced relocations from 'blackspots' (inconveniently located fragments of black-occupied land), changes in black urban housing policies which have created huge dormitory towns inside the homelands, and the judicious redrawing of white urban boundaries to exclude black townships. Mechanization of agriculture has also been a factor, but whereas this process has elsewhere forced landless labourers into urban areas, in South Africa they have often ended up in resettlement areas in the homelands.

Table 2.8 Distribution of Black population in South Africa 1960–89

	Homelands number (million)	%	'White' South Africa number (million)	%	Total Black population (million)
Year					
1960	4.1	37.6	6.8	62.4	11.0
1970	7.0	47.0	8.1	53.0	15.1
1980	11.3	54.0	9.7	46.0	21.0
1985	14.1	57.8	10.3	42.2	24.4
1989	16.1	58.7	11.3	41.3	27.4

Sources: Census data for 1960, 1970, 1980. Both the 1980 and 1985 censuses are acknowledged to have significantly underestimated the Black population, and the 1985 data are derived from adjusted data (van Zyl 1988). The 1989 estimates are from Steenkamp (1989).

Another new feature of South Africa's economic geography has been the development of border industry areas adjoining some parts of the homelands, and industrial growth points within the homelands. As already suggested, there are many economic contradictions in apartheid policies and these are another example of spatial developments running counter to theoretical expectations. The border industry and homeland growth point policies were essentially designed to shift the growth of Black employment (and therefore Black urbanization) away from the major metropolitan areas, and into the homelands. There is an extensive literature on the general failure of these policies to effect the degree of decentralization planned, largely because the comparative advantage of the white metropoles in terms of skilled labour, social and economic infrastructure, transport, and industrial linkages was too great (see for example, Mabin and Bisheuwel 1976; Rogerson 1982). Nevertheless the government implemented an impressive array of legal restrictions and inducements, with significant direct and indirect costs for the economy, which have led to the emergence of some new industrial nodes within and adjacent to the homelands. For certain types of industry, such as the textile industry which has a high demand for unskilled labour, this shift has partly been in accordance with natural economic forces (Wellings and Black 1986; Rogerson and Kobben 1982). In addition, the subsidies available in some of the homelands are so high, that returns on investment are almost inevitable. For example, in the Ciskei, there has been significant Taiwanese investment at Sada growth point, where wage subsidies could lead to a profitable return even if nothing was produced.

A further unique spatial development has been extraordinarily rapid urbanization in the homelands. From a virtually non-existent urban base in 1960, the urban population grew at 33 per cent per year in that decade, and has continued to expand since. This has been caused by virtually forced migration from rural poverty within the homelands, forced relocations from outside, dormitory development for commuters to border industries or other urban employment within range, and growth point and homeland capital city development. It is impossible to estimate the current homeland urban population for a number of reasons. One is that there is a shortage of accurate data. There is also a major definitional problem. Many settlements have urban densities, but no urban functions, particularly when they originated as dumping grounds for relocated people (de V. Graaff 1987; Unterhalter 1987). These areas have been termed 'rural slums' (Murray 1987) and include Winterveld in Bophutatswana, and Botshabelo (outside the homeland and part of South Africa). There are also sprawling unplanned settlements around homeland towns, as for example around QwaQwa's capital of Phuthudijhaba. In economic terms many residents are functionally part of white urban South Africa, since they are reliant on commuter incomes or migrant remittances. The less fortunate are reliant on charity.

Since the pass laws have been dropped and people can now seek work in white urban areas, migration to these areas has naturally increased, and squatter settlements are now a common feature within white South Africa. The population of Durban's metropolitan area is estimated to have roughly doubled over the past five years to 3.5 million, with 62 per cent of the Black population living in informal settlements.[19] The urban authorities are still trying to limit Black urbanization via housing and slum legislation, and squatter clearances are common, but as elsewhere in the world it appears that such policies cannot stem the natural economic tide.

ISSUES OF POPULATION, RACE AND ETHNICITY

The African population in southern Africa has been growing at rates of about 3 per cent per year since the 1960s. It is common to attribute many of the problems of Third World countries to rapid population growth, but within this region, the impact of population growth *per se* is very varied. By any standards, Angola, Mozambique and Zambia still have small populations in relation to their natural resource base. Lesotho and Malawi do face significant problems related to population size and growth rates. In Zimbabwe, Namibia and South Africa distortions in the distribution of wealth and resources between races have created artificially high levels of population pressure in some African rural areas, which rapid growth exacerbates. The main causes of rapid growth have been reductions in the crude death rate over the past thirty years, and continuing high fertility levels. Mortality and morbidity levels in Malawi however are still high by African standards, as are infant mortality rates in Zambia. It is possible that the health situation in both countries has worsened recently, because of economic austerity measures. During the colonial period, health conditions and infant mortality rates in the Portuguese colonies were generally extremely poor. There was evidence of rapid improvement in the early years of independence as primary health care was developed, but there is no doubt that today mortality rates are very high. It is very probable that the population growth rate in these two countries is significantly lower than the African average.

There is mounting evidence that expansion of primary health care in Zimbabwe since independence has achieved a remarkable fall in infant mortality rates from around 100 to about 50 per thousand. Furthermore this country has the highest rate of contraceptive use in sub-Saharan Africa, which may lead to further falls in fertility rates which are already slightly lower than elsewhere (although it should be noted that this will not *necessarily* follow). Some evidence suggests that a similar process is under way in Botswana, Lesotho and South Africa.

There are no reliable data on the long-term impact of AIDS on population growth rates in the region. There is however much local survey evidence to suggest that it is a serious problem. Current surveys show Malawi, Zambia and Zimbabwe to be worst affected (Akeroyd 1990), but the picture is very uncertain. At present there are far worse health problems in the region, including malaria, bilharzia and malnutrition, but the potential economic impact of AIDS is particularly serious since it is concentrated in the working-age population.

THE GEOGRAPHY OF IGNORANCE

A fundamental element of geography is the description and explanation of population distribution. The turn of a decade is always a difficult time for this task, since the last census was often ten years ago, and the latest census, if it has been undertaken at all, is not yet published. Migration patterns and urbanization levels can be estimated therefore only on the basis of past experience, official and unofficial reports and surveys, and projections. The latter in particular can be misleading if there have been major changes in economic patterns. Political instability in southern Africa makes it even more difficult to assess current population distribution. For example in Angola and Mozambique the wars have prevented official censuses, and there is a dearth of accurate information on urban populations, and the rate of population growth. It is known however that millions of people have been displaced, within and outside the countries. The last census in Zimbabwe was in 1982, and therefore an accurate picture of the impact of the post-independence era must await the next census in 1992. Similarly, in Namibia, there are as yet only the broadest estimates of the size, racial composition, growth rate and distribution of the population.

In South Africa two censuses have been taken in the past decade: one in 1980 and another in 1985. The 1985 census was taken because the quality of the 1980 data was suspect (Steenkamp 1989). However, even the 1985 census was a significant under-enumeration, and there have been valiant attempts to overcome this problem by statistical manipulation using various other population estimates (van Zyl 1988). Under-enumeration of the total population is estimated at 13 per cent, and of the Black population at 14 per cent. Within white South Africa the Black population was under-enumerated by 17 per cent, nearly one in five. In some white urban areas it is believed that almost a third of the Black population was missed. Given that there have undoubtedly been major movements in the Black population in the past five years, it is evident that knowledge of population distribution is becoming increasingly uncertain. Further efforts were made to estimate population distribution by race in 1989 (Steenkamp 1989). The statisticians involved in the 1985 and 1989 estimates both believe that many local authorities exaggerated the size of the Black urban population in their localities, and rejected some of their estimates. Whilst it seems probable that the sudden, recent influxes may have led to some overestimation, the 1989 estimates made in the reports cited here for some of the Black townships in white South Africa do not tally with many other reports of the rate of influx. In fact the 1989 report shows yet a further increase in the proportion of Blacks in the homelands to 58.7 per cent from 57.8 per cent in 1985. This simply does not fit with the growth of squatter settlements round white cities, unless all squatters are coming from other areas within white South Africa (which seems unlikely). There are therefore many important questions which need addressing: at present planners are faced with conflicting evidence.

The question of whether changing economic circumstances for the urban poor in countries like Zambia and Malawi are inducing reverse migration to the rural areas was raised in the section on urbanization. The impact of economic decline and IMF-imposed austerity measures on urban nutrition, infant mortality rates and migration all need further research. A clearer picture of the fate of rural producers, who are sometimes receiving higher prices for their products, but are also facing strong inflation in input prices, is also necessary.

The causes and effects of environmental degradation throughout the region also need further research, although much is now being done. The strength of various vested interests in land allocation however makes it difficult for some of this research to get a fair hearing. Academic objectivity is also hard to maintain in this debate, since land reform will always be a topic subject to intense ideological pressures. It is clear, however, that a narrow focus on peasant agricultural practices, which takes no note of the wider political economy within which they occur, can never lead to long-term solutions which combine justice and economic efficiency. If these two aspects are not addressed together the result may be political and economic instability – the very problems which have bedevilled the region for the past twenty-five years. Above all else, the southern African region needs stability.

NOTES

1 Malawi, Zambia, Lesotho and Botswana were formerly known as Nyasaland, Northern Rhodesia, Basutoland and Bechuanaland respectively.

2 Zimbabwe was known as Southern Rhodesia until UDI in 1965, as Rhodesia until 1979, and as Zimbabwe-Rhodesia until independence in 1980.

3 MPLA is the Movimente Populare para Liberacao da Angola (Popular Movement for the Liberation of Angola); FRELIMO is the Frente para Liberacao da Moçambique (Front for the Liberation of Mozambique).

4 Strong political pressure from their supporters, Mozambique and Zambia, which were experiencing a strategic escalation in military and economic pressure from the South Africans and Rhodesians at the time,

also played a part in the PF's agreement to the Lancaster House constitution.

5 The Mozambique National Resistance is also called RENAMO (Resistencia Naçional da Moçambique).

6 UNITA is the Union Nacional para Independencia Total da Angola (National Union for the Total Independence of Angola).

7 Neither Swaziland nor Malawi were perceived as threats to South Africa's regional objectives.

8 The degree of political resistance to these programmes has varied. Malawi has accepted a series of structural adjustment programmes, and many, although not all, the conditions have been met. Zambia began a programme in 1986, but after food riots in Lusaka, swiftly ended it. In 1990 another programme was adopted as Zambia's economic plight deepened.

9 This programme was known as the SEF (Saneamento Economico e Financeiro or Economic and Financial 'cleaning'). Angola's independent adoption of reforms, rather than an IMF programme with external financial backing, was partly enforced since the USA prevented its membership of the IMF until 1989.

10 The most important movement is the African National Congress (ANC), and its close ally, the South African Communist Party (SACP). To what extent the SACP's ideology influences the ANC's is naturally a matter of intense debate. Mandela's public statements suggest an ideology far less radical than communism, but the situation is dynamic and it is impossible to be definitive. It is also important to recognize that other ideological positions are maintained by the other liberation movements, which include the Pan-African Congress (PAC), Azanian People's Organization (AZAPO), and the Black Consciousness Movement (BCM). Trade unions, particularly the hugely influential Congress of South African Trade Unions (COSATU), are also significant actors on the political scene. Inkatha, the Zulu-based organization led by Gatsha Buthelezi, is very powerful and at ideological odds with all of the above. The dynamics of the current South African political scene are therefore very complex. See for example Suckling and White 1988; Johnston 1988; Hirschmann 1990; Southall 1987.

11 Malawi's infant mortality rate is probably the highest in the region, with the exception of the war-torn Angola and Mozambique, for which there are no data. Figures published by the usual agencies, such as the World Bank and UN, bear little relation to survey data. An unpublished World Bank sponsored survey in the early 1980s for example found rates of 300 were not uncommon in the Central Region. Zambia's rates may however be rising again, particularly in the urban areas.

12 There is an extensive literature on this topic. See, for example, Palmer and Parsons 1977; Bohning 1981; Murray 1981; Low 1986; First 1983.

13 The communal areas is the new term for the rural areas which were designated for African occupation under various land acts prior to independence. These areas were originally known as the native reserves, and then after UDI as the Tribal Trust Lands.

14 Other important considerations for the overall success of peasant

agriculture in resettlement areas and elsewhere, but which do not have geographical implications *per se*, are the political and social organization of rural communities.

15 The fostering of trade between African states in southern and eastern Africa is covered by another, quite separate organization, the PTA (Preferential Trade Area).

16 These are: Angola – energy; Botswana – crop and veterinary research; Lesotho – soil conservation; Malawi – fisheries and wildlife; Mozambique – transport; Swaziland – manpower development; Tanzania – industry; Zambia – mining; Zimbabwe – food security.

17 Racial nomenclature in South Africa is very specific. Four basic racial groups are recognized for population registration purposes: White, Black, Asian and Coloured. The term 'black' is used for all non-whites, whilst the term 'Black' is used for the indigenous African people who speak Bantu languages.

18 In recent years the government has supported the establishment of the Southern African Development Bank which has taken over much of the direct financial support necessary to maintain homeland governments.

19 These estimates were given to me by the Built Environment Support Group at the University of Natal, Durban, in August 1990.

REFERENCES

'African Business' (1989) *Zimbabwe Survey*, May.

Akeroyd, A.V. (1990) 'Aids in southern Africa: an overview', paper given at African Studies Association UK biennial conference, University of Birmingham, mimeo, September.

Bohning, W.R. (ed.) (1981) *Black Migration to South Africa*, Geneva: International Labour Organization.

Browett, J.G. and Fair, T.J.D. (1974) 'South Africa 1870–1970: a view of the spatial system', *South African Geographical Journal* 56: 111–20.

Central Statistical Office (1989) *Quarterly Digest of Statistics: September 1989*, Harare: Central Statistical Office.

Christiansen, R.E. and Kydd, J.G. (1983) 'Return of Malawian labour from South Africa and Zimbabwe', *Journal of Modern African Studies* 21 (2): 311–26.

Christiansen, R.E. (1984) 'The pattern of internal migration in response to structural change in the economy of Malawi 1966–77', *Development and Change* 15: 125–51

Davies, R.J. and Cook, G.P. (1968) 'Reappraisal of the South African urban hierarchy', *South African Geographical Journal* 50: 116–21.

Fair, T.J.D. (1982) *South Africa: Spatial Frameworks for Development*, South African Geographical and Environmental Series, Cape Town: Juta.

Fauvet, P. (1979) 'Education in Angola', *People's Power* 15: 41–9.

Financial Times (1981) *Zimbabwe Survey*, i–viii.

Financial Times (1989) *Zimbabwe Survey*, 21 August: i–viii.

Financial Times (1990) *Namibia Survey*, 22 March: 17–20.

First, R. (1983) *Black Gold: The Mozambican Miner*, Brighton: Harvester Press.

Fransman, M. (ed.) (1982) *Industry and Accumulation in Africa*, London: Heinemann.

Frederikse, J. (1982) *None but Ourselves: Masses versus Media in the Making of Zimbabwe*, Harare: Zimbabwe Publishing House.

Geldenhuys, D. (1981) *The Constellation of Southern African States and SADCC: Towards a New Regional Stalemate?*, Braamfontein: South African Institute of International Affairs.

Graaff, J.F. de V. (1987) 'The present state of urbanisation in the homelands: rethinking the concepts and predicting the future', *Development Southern Africa* 4 (1): 46–66.

Hall, M. (1990) 'The MNR (Renamo): a study in the destruction of an African country', *Africa* 60 (4): 39–68.

Hanlon, J. (1986) *Beggar Your Neighbours: Apartheid Power in Southern Africa*, London: Catholic Institute for International Relations.

Hirschmann, D. (1990) 'Of monsters and devils, analyses and alternatives: changing Black South African perceptions of capitalism and socialism', *African Affairs* 89 (356): 341–70.

Johnston, S. (ed.) (1988) *South Africa: No Turning Back*, London: Macmillan.

Kydd, J. (1989) 'Maize research in Malawi: lessons from failure', *Journal of International Development* 1 (1): 112–44.

Kydd, J. and Christiansen, R. (1982) 'Structural change in Malawi since independence: consequences of a development strategy based on large-scale agriculture', *World Development* 10 (5): 355–75.

Lipton, M. (1988) 'South Africa's role in agricultural production and marketing in Southern Africa', in Bryant, C. (ed.) *Poverty, Policy and Food Security in Southern Africa*, London: Mansell.

Low, A. (1986) *Agricultural Development in Southern Africa: Farm Household Economics and the Food Crisis*, London: James Currey, and Cape Town: David Philip.

Mabin, A. and Bisheuwel, S. (1976) 'Spatial implications of apartheid: role of growth centres in South Africa', *Pan-African Journal* 9 (1): 1–16.

Minter, W. and Schmidt, S. (1988) 'When sanctions worked: the case of Rhodesia re-examined', *African Affairs* 87 (347): 207–38.

Murray, C. (1981) *Families Divided: The Impact of Migrant Labour in Lesotho*, Cambridge: Cambridge University Press.

Murray, C. (1987) 'Displaced urbanization: South Africa's rural slums', *African Affairs* 86 (344): 311–29.

Nelson, H. (ed.) (1987) *Mozambique: A Country Study*, Area Handbook Series, Washington DC: American University.

O'Connor, A. (1983) *The African City*, London: Hutchinson.

Palmer, R. (1990) 'Land reform in Zimbabwe: 1980–1990', *African Affairs* 89 (355): 163–82.

Palmer, R. and Parsons, N. (1977) *The Roots of Rural Poverty in Central and Southern Africa*, London: Heinemann.

Peet, R. (1984) *Manufacturing Industry and Economic Development in SADCC Countries*, Stockholm and Uppsala: Beijer Institute and Scandinavian Institute for African Studies.

Pottier, J. (1988) *Migrants No More: Settlement and Survival in Mambwe Villages, Zambia*, London: Manchester University Press.

Potts, D.H. and Mutambirwa, C.C.M. (1990) 'Rural–urban linkages in contemporary Harare: why migrants need their land', *Journal of Southern African Studies* 16 (2): 676–98.

Rogerson, C.M. (1982) 'Apartheid, decentralisation and spatial industrial change', in Smith, D.M. (ed.) *Living Under Apartheid*, London: Allen & Unwin.

Rogerson, C.M. and Kobben, S.M. (1982) 'The locational impact of the EPA on the clothing and textile industry', *South African Geographer* 10: 19–32.

Shaw, J. (1981) 'Functional co-operation in Southern Africa: an experiment in consensus decision-making', *Development Studies on Southern Africa* 4 (1): 2–34.

Smith, D. (1985) *Update: Apartheid in South Africa*, Cambridge: Cambridge University Press.

Smith, S. (1990) *Frontline Africa: The Right to a Future*, Oxford: Oxfam.

Southall, R. (1987) 'Post-apartheid South Africa: constraints on socialism', *Journal of Modern African Studies* 25 (2): 345–74.

Steenkamp, H. (1989) *Regional Population Estimates for 1989*, Pretoria, University of Pretoria: Bureau of Market Research.

Stoneman, C. (1989, 1990) *Zimbabwe Quarterly Surveys*, London: Economics Intelligence Unit.

Stoneman, C. and Cliffe, L. (1989) *Zimbabwe: Politics, Economy and Society*, London: Pinter.

Suckling, J. and White, L. (eds) (1988) *After Apartheid: Renewal of the South African Economy*, London: James Currey.

Unterhalter, E. (1987) *Forced Removal: The Division, Segregation and Control of the People of South Africa*, London: IDAF.

Wellings, P. and Black, A.M. (1986) 'Industrial decentralization in South Africa: tool of apartheid or spontaneous restructuring?', *Geojournal* 12 (1): 137–49.

World Bank (1989) *Sub-Saharan Africa: From Crisis to Sustainable Growth*, Washington DC: World Bank.

World Bank (1990) *World Development Report 1990*, Washington DC: World Bank.

Zinyama, L. (1990) 'Agricultural development policies in the African farming areas of Zimbabwe', *Geography* 311 (71): 105–15.

Zyl, J. van (1988) *Adjustment of the 1985 census population of the Republic of South Africa by district*, Pretoria, University of South Africa: Bureau of Market Research.

3

THE CHANGING GEOGRAPHY OF CENTRAL AFRICA

Felicité Awassi Atsimadja

INTRODUCTION

The countries reviewed in this section on Central African States are Cameroon, The Central African Republic (CAR), Chad, Equatorial Guinea, Burundi, Rwanda, the Popular Republic of Congo (PRC), the Democratic Republic of Gabon (DRG) and Zaire (see Figure 3.1). This group is conventionally known as Central Africa (except that Burundi and Rwanda are sometimes classified with East Africa), but it is not a uniform group – Chad for example stretches from the Sahara desert southwards into savannah. Zaire is mostly within the tropical rain forest belt, and Rwanda and Burundi are both hilly and mountainous countries. They also vary significantly by size – both geographically and in population, and by density of settlement (see Table 3.1). Rwanda and Burundi are small in terms of population and geographical area, but high in population density. Zaire has the largest population in this group, yet because of its vast geographical size (four times that of France) is settled at a very low density. They vary greatly in terms of resource endowment too. Zaire has a wealth of minerals, including copper and uranium. Gabon was poor at independence, but soon after struck a major oil field, which has transformed the state's finances. Cameroon is expanding its smaller oil industry. The Central African Republic has a wealth of diamonds, and also valuable metals. On the other hand, Equatorial Guinea, Rwanda, Burundi and Chad have poor mineral endowments.

If there is any unity in this diverse group it is found in three aspects: first, this is the 'heart' of Africa, and its centrality and remoteness meant that these were the areas last penetrated by

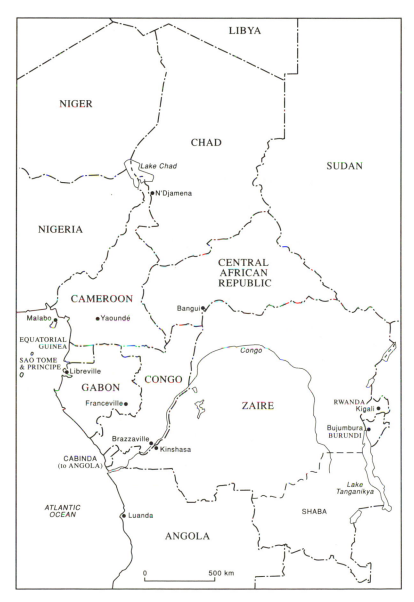

Figure 3.1 Central Africa

53

Table 3.1 Basic population parameters in Central Africa

	Population (million) 1960	1990	% Annual population increase 1984–9	Area '000 sq.km	Density persons per sq.km	% Urban population 1990	Infant mortality 1970	1990	Life expectation Male	Female
Burundi	2.91	5.44	2.9	27.8	195.7	8.2	150	114	46.9	50.2
Cameroon	5.61	11.74	2.9	456.4	25.7	28.5	137	94	49.0	53.0
CAR	1.5	2.88	1.8	622.4	4.6	27.4		132	43.9	47.1
Chad	3.65	5.68	2.5	1284	4.4	16.0	190	132	43.9	47.1
Congo	0.96	2.33	3.6	342	6.8	51.1	160	124[a]	46.9	50.2
Equatorial Guinea	0.24	0.35	2.3	28.1	12.5	27.6	180	127	44.9	48.1
Gabon	0.63	1.27	1.9	267.7	4.7	40.9	229	103	49.9	53.2
Rwanda	2.74	7.23	3.4	26.3	274.9	4.5	137	122	46.9	50.2
Zaire	16.15	34.14	2.4	2345.1	14.6	36.2	104	98	50.8	54.2
	34.39	71.06								

Source: All Encyclopaedia Britannica Yearbook (1990), except 'Infant mortality' from UN Statistical Yearbook (1970) cited in Ochola (1975), and [a] from The Courier (1991) vol 125.

Table 3.2 Employment and sectoral composition of GDP in Central Africa

	% Land area cultivated (excl. pasture)	% Employment Agriculture 1970[a]	% Employment Agriculture 1986[a]	% Employment in Industry 1986[b]	% Sectoral origin of GDP 1986[b] Agriculture	Mining	Manufacturing	Construction	Transport	Financial
Burundi	85.8	86.0	93.1	1.7	63	1	10	5	9	11
Cameroon	3.3	82.0	74.0	4.6	25	17	11	7	33	8
CAR	0.8	87.0	67.8	4.2	41	2	7	3	4	20
Chad	45.8	91.0	81.2	5.2	46	1	9	2	(total 42)	
Congo	0.7[c]	45.0	61.2	12.1	12	16	10	7	27	28
Equatorial Guinea	5.7	na	85.7	0.9	59		1	11	8	21
Gabon	0.3	72.0	73.5	11.1	7	24	5	11	18	26
Rwanda	51.3	91.0	92.9	1.8	40		16	7	17	21
Zaire	2.6	78.0	68.1	14.5	32	25	2	5	20	15

Sources: [a] Production Yearbook, FAO, cited in Ochola (1975).
[b] Encyclopaedia Britannica Yearbook (1990).
[c] The Courier (1991) vol 125 gives 2.0%.

European powers; second, because (apart from a small part of Cameroon and tiny Equatorial Guinea) they were colonized by the French and Belgians, there is a *lingua franca* for the educated – French; and third, much of the area (in Congo, Zaire, Central African Republic) shares a major common resource – the Zaire (Congo) River basin.

The combined area of these countries is vast – 5.4 million square kilometres, but in this area the total current population is only about 72 million. The region as a whole is, remarkably, the most urbanized of all the regional groupings of sub-Saharan Africa – approaching 40 per cent. These figures combine to mean that settlement density, particularly of the rural population, is very low. The average amount of land cultivated in the region is only 13 per cent (see Table 3.2); much does still remain tropical forest and savannah in good if not pristine health. There are significant wild-life reserves and national parks – which could constitute a major resource, particularly for tourism, although some of these resources are at risk. But the underdevelopment of transport links, and the fact that the major rivers are not navigable from the sea means that there is some inherent degree of isolation and protection.

The Central African Republic, Congo, Chad, Cameroon (see p. 59 for a caveat) and Gabon were all formerly part of the French Equatorial African Empire. Zaire, formerly the Belgian Congo, was a private colony of the Belgian king for a while in the nineteenth century, but became administered by the Belgium government and parliament. Rwanda and Burundi were also Belgian colonies. Equatorial Guinea was a Spanish colony, but has allied itself economically more closely in recent years with the former French colonies. Since 1985 it has joined them in using the CFA franc as its currency. This is tied to the French franc and is convertible. Zaire, Rwanda and Burundi have their own currencies. The Central African Republic, Gabon and Congo are linked together in UDEAC (Union Douanière et Economique de l'Afrique Centrale) – which Chad might also rejoin.

Since independence the 'performance' of these states in crude aggregate terms (the value of which will be disputed in the section on Industry, pp. 70–2) has varied greatly. As O'Connor (1991) has shown, over the fifteen years 1973–88, average income per capita went up strongly only in Congo, Rwanda, Burundi and Cameroon; it went up slightly in Gabon; but in Equatorial Guinea and Zaire it actually declined (Table 3.3). Gabon was 'born' one of the poorer

Table 3.3 Economic growth in Central Africa

	p.c. GNP 1989 US$	% Annual growth real GNP	
		1970–80	1980–7
Burundi	260	4.0	3.9
Cameroon	900	4.5	6.5
Central African Republic	215	2.6	2.1
Chad	90	0.5	4.2
Congo	1300	4.7	6.1
Equatorial Guinea	260	−9.4	−3.9
Gabon	2800	7.5	−1.0
Rwanda	290	3.0	2.1
Zaire	190	0.5	1.7

Source: World Today: Africa (1989).

countries of the region, but then struck oil, literally and consequently metaphorically too. The value of this oil has of course
fluctuated enormously, but it is still better to be with it than
without it. The wealth has enabled the government to tackle
some schemes like the new Libreville–Franceville railway with
confidence, but it clearly has to be careful about how many projects
to start when future income is not assured. So, in interpreting the
statistics for income change one has to remember that recent
declines for Gabon have much to do with the height of previous
peaks, such as under the first (1973) and second (1979) OPEC oil
shocks. Fluctuating prices also have a major impact on other
mineral industries too: copper, so important to Zaire, has dropped
heavily in price in the 1980s. However, the changing fortunes of
these states has not been due solely to fluctuating external commodity prices: it has had as much to do with political and
governmental stability and competence.

THE POLITICS OF INDEPENDENCE AND INDEPENDENT DEVELOPMENT

During the early 1960s as the majority of African countries south
of the Sahara were gaining their political independence, some
Africanists were already questioning whether these countries would
be able to achieve independent economic development and what in
truth political independence would mean. This pessimism (see
Chapter 1 of this volume) was best voiced by René Dumont (1962)
whose main conclusion was that there had been a 'false start' in

black Africa, because Africa had started its independence in adverse conditions and would therefore now collapse. The main reason for saying this was that the overhead capital necessary for economic take-off did not exist. The African governments had therefore the arduous task not only of constructing their basic infrastructure, but also of finding the best path towards development, when perhaps many of the existing models of development were not appropriate to these exacting conditions. The governments therefore in a sense had to act in a policy vacuum. A few years later predictions from the West were even gloomier. Meister (1966) thought that the real question that ought to have been posed at independence was not whether Africa could develop but whether Africa could even begin to think of starting the process of change *on its own*, despite its independence.

However, the outlook on African prospects during the first decade of political independence was not exclusively pessimistic. African leaders in the main were impelled by their new responsibilities to be optimistic. This confidence in the future was expressed by President Bongo of Gabon whose favourite motto in the second half of the 1960s was 'faîtes moi de la bonne économie et je vous ferai de la bonne politique' ('Give me a good economy and I will give you good politics'). There is a considerable irony in this remark. Although in some ways the economic performance of much of Central Africa over the last two decades or so has been better than that of many other parts of Africa, there have been severe set-backs, and the region as a whole has performed nowhere near as well as it might have done. The poor economic performance has often (but not exclusively) been allied with poor political performance – a combination of unstable and/or ruthless governments, inappropriate and effectively inegalitarian policies (although often explained in egalitarian terms) and internal disorder. For many observers, President Bongo should have reversed his priorities: the deeper question is why has the achievement of 'good politics' been so difficult. This touches on a theme to which we will return several times in this chapter.

Before European penetration and domination of the heart of Africa, only a little over one hundred years ago, social and territorial organization bore no resemblance to the European concept of sovereign territorial states. Settlement densities were, as they still are, very low, thus land *per se* was rarely an issue to be contested. The indigenous people were divided into a myriad of tribes, often with separate languages. Much of the economy was based on hunting/fishing/gathering as much as farming – which was usually shifting cultivation of some kind. Different groups often traded with each other – swapping gathered forest products for food crops for example – and there was some exploitation of one group by another in slave trading for example. But there was little or no wider over-arching authority. European powers divided the area in a fairly arbitrary way. Many boundaries might not have seemed arbitrary in that they followed a river or a watershed, although there are some geometrical regularities that look physiographically meaningless – for example the boundaries of the mainland part of Equatorial Guinea. The problem was that the boundaries often neglected tribal divisions, and indeed were drawn on inaccurate maps by powers that often had only a poor idea of the distribution of tribal groups. But division by tribal groups would be impossible anyway – since they often intermingle, and there are too many so that in terms of population and geographical spread they are often too small to form the basis of a nation state.

Therefore although there have been many tensions over the arbitrary boundaries, the policy of the OAU that the inherited boundaries be respected has in the main been adhered to. There has been one notable exception – settled peacefully. The original German colony of Cameroon was divided under a League of Nations resolution at the end of the First World War into the two separate protectorates of British and French Cameroon. After independence, the two voted to rejoin to form one republic, except that part of the British protectorate voted to join Nigeria in 1961, and did so, thus altering the colonial boundary as originally set. There has also been one determined attempt at secession. In Shaba Province, then Katanga Province, of Zaire, a secessionist government actually ran the province as an independent country for three years, between 1960 and 1963, when a military campaign supported by United Nations troops finally squashed the rebellion. More recently the northern part of Chad was effectively beyond the authority of the government in Ndjamena. By the late 1980s the country had been reunited under force of arms, and hopefully a political accommodation can follow.

It seems then as if there is general acceptance that the states as currently defined territorial entities will persist. But such an acceptance does not mean that building national consciousness is easy, or that tensions will not arise. The northerners in Gabon for instance have family connections in Cameroon and Equatorial Guinea, while families from the Haut Ogoué, Bateke and Ambama regions of Gabon's south-east have relatives in Congo (Brazaville). How is it that a myriad of peoples split by these borders and united with dissimilar groups in a sovereign territory should form a new nation?

As is well known, independence in Central Africa was rapidly followed by the establishment of one-party states, or even one-president states, little more than personal fiefdoms. In some ways this is explicable in that the model of government that had been portrayed in all the states by the colonial powers was of strong central paternalism, whether benevolent or otherwise, and of course there had been no history of democracy. But in addition the one-party system seemed to offer a means of avoiding divisive politics based on tribal lines which would accentuate the tensions we have just noted. Such centralized systems may have achieved more in unifying these states than they have been given credit for – even if only in focusing opposition hostility against a single target. But clearly they are deficient in allowing for political evolution and change: indeed change at the top has been quite commonly through a coup, and dissent noticed by those at the top often only after a failed coup. Such failed coups have on occasion been dealt with savagely. In Burundi in 1972 President Micombero, one of the dominant Tutsi (who form only about 18 per cent of the population), assumed dictatorial powers. Opposing Hutu unsuccessfully attempted his overthrow, and in subsequent genocide some 120,000 people – nearly 3 per cent of the population – died.

Some of these coups have replaced one-party government with one-army government. Some have produced almost literally one-man government – perhaps Mobutu of Zaire has come close to that definition in the past, though his power is less absolute than in the past. The ideological nature of these governments has also varied. Some like Congo have formed hard-line Marxist-Leninist states – and only now in 1991 are Congo and Gabon bowing to the winds of change in the Comecon Bloc and attempting to find a way to a multi-party more market-orientated economy. For example in Zaire three main parties have recently emerged in addition to the previous single party MPR (Mouvement Populaire pour le Renouveau). The new ones are L'Union pour la Démocratie et le Progress Social (UDPGS); the Parti Démocrate et Social Chrétien (headed by Joseph Ileo and supported by German and Belgium Christian

Democrats); and the Lumumba MNC. In Gabon a new press freedom has spawned a number of new newspapers, where once only the official L'Union was permitted: they are Missamu (mouthpiece of Morena), La Clé, La Griffe, La Rélance and Le Couperet. The left-wing states of Africa like Angola or Congo can no longer expect assistance from Eastern Bloc countries. Economic power in world terms is increasingly centred in North America, the European Community (EC) and Japan. International institutions increasingly share a currently fashionable common philosophy which stresses the discipline of the market and the growth potential of private enterprise. And in Central Africa itself, the fruits of substantial education programmes are being felt. Here, where literacy levels are relatively high, more and more people wish to participate in the process of building their own national futures. The convergence of all these forces is plain to see: but the path to democratic government and sustainable development will not be easy, even if there does seem agreement that that is a worthwhile goal. In some of these states it would be hard even now to distinguish an ideological position, other than the empowerment and enrichment of a usually small ruling elite.

POPULATION GROWTH

During the past three decades, the rate of growth of national populations has generally been very high (see Table 3.1) if not quite at Kenya's spectacular rate. For example an annual growth rate of above 2.8 per cent implies a population doubling time of only twenty-five years, and a population quadrupling of only fifty years. Out of these eight countries, six have population growth rates that exceed 2.8 per cent.

But it is noticeable that the population totals, even after recent decades of rapid growth, are in the main small in absolute terms. The low population densities that result are often blamed for some of the problems of development: the base is too small to support large-scale infrastructure development, and market potentials even with increased per capita incomes will always stay low. Therefore at one level population growth should be welcome. But if it happens too rapidly, then of course the population pyramid becomes heavily weighted towards the bottom end, and the number of dependent children supported by the working population increases proportionately, and so this can become something of a burden on current consumption and a brake on investment.

The population growth is the result of the first classic stage of the demographic transition, when death rates are declining faster than birth rates. The fall in death rates is mostly the result of improved health facilities and improved drugs (see Table 3.4). The decline in birth rates is significant but small. In Burundi the birth rate declined from 46.1 per thousand in 1970 to 42.0 in 1980, and in Cameroon birth rates declined from 49.9 in 1970 to 42 in 1980. The pattern is repeated elsewhere, but large families are still a characteristic of all these societies, and average numbers of children per fertile woman very high. In Rwanda the birth rate per thousand is 51, a figure exceeded world-wide only by Kenya and Malawi. The women of Rwanda have been dubbed the champions of fecundity (*The Courier* 1982a) as they have the highest 'indice synthétique' with 8 children, compared with 6.5 in Burundi, 6.1 in Zaire, 4.4 in India, 2.3 in China and 1.5 in Western Europe.

Table 3.4 Health care provision in Central Africa

	Population ('000s) per physician 1965	1984	1991	Population ('000s) per nurse 1965	1984
Chad	72.4	38.4		13.6	3.4
Burundi	55.9	21.1		7.3	3.0
Rwanda	72.5	34.7		7.5	3.7
CAR	34.0	23.6		3.0	2.2
Congo[a]			5.5	1.0	0.6
Gabon	14.2	2.8		0.8	0.3

Source: World Bank (1990: 232–3)
Note: [a] 1991 figure from *The Courier* (1991) vol 125.

As in East Africa, AIDS is a problem, though in Central Africa it is not yet at the same levels as it is believed to be in Uganda. It is difficult even in developed countries to get good data because of ethical problems over privacy even when a blood sample has been taken for other reasons, and in the developing world equal ethical reluctance is exacerbated by inadequate access to much of the population and by the lack of resources to commit to caring for the sufferers. In Kinshasa, Zaire in 1988 (*The Courier* 1988b: 26) it was thought that 3–8 *per cent* of the population was sero-positive – a frightening rate if true and if also true of the rest of the country, which is by no means certain. The figure for Equateur province was 1 per cent, but 4–8 per cent in pregnant women. It was 27 per cent among Kinshasa prostitutes. The government in Zaire acknowledges the problem, and has a sustained awareness campaign, aimed partly at altering people's behaviour. Its success or otherwise is yet to be determined. Most forecasts make grim reading.

URBANIZATION

Before independence the growth of urban areas was restricted by direct planning controls, and by subtle social discrimination: much of this was for specific political purposes, to try and contain the growth of a new indigenous middle class who might challenge the European authorities. It is of course not possible to prove that in the absence of such discrimination towns would have grown faster, but it is significant that urbanization has accelerated markedly since independence. It is by no means uniform over the region: Rwanda and Burundi, densely settled farming areas, remain with a low level of urbanization – but most of the populace do have some access to urban facilities when needed. In Chad the urbanization rate is still low, and here the population, which lives at the lowest density for the whole region, has little access to urban facilities. In the other states, however, the level of urbanization has exceeded the average for most parts of Africa. In Zaire, Congo and Gabon, it is now almost as high as for some Latin American countries. For these countries this constitutes a real problem: not just of pressure on urban facilities, felt in cities throughout Asia and Latin America as well as here, but because it does not have a beneficial impact on rural areas. Whereas in some parts of rural Asia it may be said to have a good impact in rural areas (or at least a not-bad-impact, as pressure on arable land is so high), in the Central African countries rural depopulation is eroding the development capacity of the remaining rural communities. Initially they were denuded of most of their active male population – an asymmetry partly induced by colonial legislation admitting male workers to specific jobs, but excluding the dependants – a practice still continued in South Africa under apartheid. This has left old people and children at risk, with less adequate support. Recently the pattern in some areas of Central Africa has been much more one of whole family migration. In this sense the problems of rural depopulation are more akin to the problems of peripheral rural communities in Europe than they are of many Third World areas.

To take an example, in Cameroon the average rate of growth of the urban population has been around 6 per cent, implying a

63

doubling of the urban population in less than twelve years, less than half the doubling time for the population as a whole. In 1970 the country was 18 per cent urban: but now it is 28 per cent. The urban population is not evenly distributed – the coastal regions are far more urbanized than the interior regions – so the advantages of a nearby urban centre are not well distributed throughout the country. The unevenness of urban development is characteristic of most of the other states too.

The reasons for these patterns are many and complex. There may however be some general factors at work. The colonial societies developed health and educational infrastructure disproportionately in the 'major' urban centres. This lop-sided provision has not yet been corrected, and urban areas are seen as offering far greater security, particularly in health, than rural areas. As Lassère (1970) put it,

> It is certain that for the man in the countryside, the town is a place where he feels more secure. ... In Gabon, in Libreville, among the factors that most led to immigration, this feeling of security figured prominently as the most common cause of immigration. This was particularly true for women; the existence of a maternity hospital was an extremely attractive factor. For instance in a Libreville quarter which bears the significant name of 'Derrière L'Hôpital', most neo-urban dwellers settled around that area in order to be as near as possible to the Hospital.

Differences in child and adult mortality rates between urban and rural areas tend to bear out this perception, and hence if the move to an urban area is economically feasible, it is also rational. And indeed economic motivation was also provided by these countries, in that for some years after independence stress was put on the industrialization of the economy, processing imported raw materials and semi-finished goods, or processing for exports. Price discrimination against agriculture also eliminated incentives for rural communities.

Capital cities particularly experienced a huge increase in their population, and in all these countries the degree of primacy of the largest city has increased. It is also noticeable that such concentrated urbanization is another way in which the population becomes increasingly unequally spread between regions. In Congo,

Brazzaville and Pointe Noire on the coast, plus a small strip of land around the rail and road links in between, account for half of the country's population. In Chad the population is overwhelmingly southern – the five south-western prefectures have 10 per cent of the area and 47 per cent of the population. Ndjamena may look somewhat eccentricly placed within the national territory, but is in fact much more central with regard to economic activity. In Zaire and Congo the two capital cities of Brazzaville and Kinshasa have an overwhelming primacy – and the two, as twin cities on the banks of the lower Congo, make a new heart in Africa.

The new urban cultures

Urbanization is the process by which an increasing percentage of a country's population becomes resident in urban areas. Urbanism is a way of life – distinct in attitudes and values from the rural ways of life and the myopia of the small farmer. The two are linked in the way that, for example, there are more schools, particularly at higher levels, in towns than in rural areas, and schools are important in inculcating new attitudes and values.

Schooling enrolment rates rose in Gabon between the 1960s and 1990s. They went from the already very high 67 per cent of the 1960s to 80 per cent in the 1970s and by 1980, had reached 100 per cent. One obvious and logical consequence of this has been a shift from a male:female pupil ratio of 66:33 to a balanced 50:50. Although Gabon's spectacular case can be explained in part through oil-wealth, this is not the explanation for the generally high rates of schooling in many other Central African countries. By 1977, schooling rates in most Central African countries had gone beyond the 50 per cent level: Cameroon 80 per cent, Gabon 90 per cent, CAR between 60 and 65 per cent, Congo 95 per cent, Zaire 85 per cent, Rwanda 76 per cent. Only Burundi and Chad languished behind with schooling rates of 24 per cent and 30 per cent respectively. These figures tend to suggest that the service-access thesis of urbanization is tenable: Rwanda and Chad are in the three least urbanized states of the region. Only rural Burundi, the second least urban, has bucked this particular trend.

It is among the youth of the expanding towns that the search for a new national or pan-national identity is strongest. In one way change is partly the result of the kind of international homogenization that Western urban technology and culture engenders. To that extent the modern urban population, better educated and probably able to speak in French, represents a detraction from the process of building purely national states. This modern youth has its own vehicles of expression. The music which emanates from the clubs and bars of Kinshasa and Brazzavile is the strongest expression known outside of Africa, apart possibly from the pride felt beyond the Cameroons in that country's performance in the soccer World Cup of 1990. A transnational formal attempt at preserving and projecting Bantu culture has led to the formation the Centre International de Civilisation Bantu (CICiBA) in 1983. Its ten members, Angola, CAR, Comores, Congo, Gabon, Equatorial Guinea, Rwanda, Sao Tomé and Principe, Zaire and Zambia include most of these Central African states.

On the other hand, a search for indigenous authenticity can stress tribal origins. In Zaire President Mobutu, who has successfully held power since 1965 when as a general he seized power, has promoted a return to African cultural origins, by, amongst other things, banning European personal and geographical names. He himself changed from Joseph Désiré Mobutu to Mobutu Sese Seko. It is difficult to know the extent to which such a move will accentuate older tribalism – but it is noticeable that since opposition parties were allowed to form for the first time in Zaire in 1990, more than eighty have been registered.

Abdoulaye Wade (1990) likens Africa's present plight to that of a France that would be divided into twenty small states, or the USA when broken into several mini states and concludes with their inevitable non-development.

> Africans and in particular the Continent's young generations believe that none of the African States will be able to develop within the confines of their present individual geographical spatial boundaries. Africa is being led towards the opposite direction in which other continents are engaged.

Abdulaye Wade also believes that despite the initial euphoria born out of the novelty of democracy, which saw a mushrooming of political parties (seventy-four political associations in Gabon in 1990), political expression will soon be reduced to bipartism as in the USA or a three-party system as in Germany. (This is a difficult prophecy to defend.)

ECONOMIC CHANGE IN CENTRAL AFRICA

Communications

This part of Africa is above all about communications: their essential necessity and equally their virtual absence. Chad's capital Ndjamena is by any route at least 1,500 kilometres from the sea. In Central Africa the Congo basin is traversed by thousands of kilometres of navigable waterways, but is nevertheless cut off from the sea. (The Great Lakes of North America are cut off from the St Lawrence seaway by the Niagara Falls, but it has been possible to circumnavigate the latter with great locks. It has not been possible in the case of the Congo, and it is highly doubtful whether it ever will.) The link between Brazzaville and Kinshasa and the sea is by road, and by difficult and expensive railways in both cases. Ironically, they themselves, twin cities almost facing each other on the river, are still not connected by either a road or a rail bridge although EC plans are being advanced in 1991 for funding a road–rail connection. However, completion by the turn of the century is unlikely.

There are of course national and international plans to improve communications. The most ambitious scheme is underway, to drive a pan-African highway from Lagos to Mombasa, right across the Zairean bush and jungle. Most schemes are however small in relation to the overall task, and always with few resources available. Gabon's area is slightly greater than that of the United Kingdom, yet it has even now only 1.25 million people. In 1965, Gabon's overhead capital was virtually non-existent. There were no tarred roads. The few dirt roads were impassable during the rainy season. The government saw it as a priority to improve communications, and amongst other moves established Transgabonais in 1973 as a national company responsible for the development of the railways. Between 1970 and 1980 direct road links were established between various areas of the countries which had been connected only by air link or not at all. These were the Libreville–Franceville routes, the Makokou Akondja routes, the Lalara Koumameyon and the Lalara Mitzic routes in the northern parts of the country. Laterite roads were built between Franceville and Leyou, between Kelle and Lekoni and Ndjole and Mevang. By 1980, Gabon had 6,929 km of high-grade public roads, 459 km of which were tarred. The impact of the new system on accessibility has been dramatic within the limited areas served. For instance before 1965, it took four days to go from Libreville to Mimongo by land and it took three days to travel from Libreville to Oyen. Today these journeys take a few hours. Nor is it just urban–urban communications that have benefited. Now many rural areas have access to urban outlets for their produce. They have an incentive to produce commercially, and the country's import bill for food-stuffs can decline a small amount. Transgabonais has developed a new rail link between Libreville and Franceville, to help in exploiting the forests and the uranium deposits of the interior. The project was put forward to the World Bank for financing, but it was rejected as too expensive and not cost-effective. Feeling confident in the wake of rising oil prices, the government started to build its railway in 1974 by borrowing on the open market (*The Courier* 1991). The costs of blasting, and of erosion control, rapidly showed themselves to be far higher than the initial estimates – five times higher in the case of rock excavation. The line to Franceville was completed in 1986, and is now carrying a greatly increased freight traffic between the interior and the coast. But, like the lines built in the nineteenth century by the British to the Kenyan highlands or the Nigerian plateau, it is almost certain the the traffic will never repay the capital and interest. It will remain a significant drain on the public purse for many years to come. It is also ironic that it has been built at the standard international gauge, whereas the systems to which it ought possibly to connect in Congo and Zaire in the future are at 1 metre gauge.

The climb by the railways in Congo and Zaire from the sea through the mountainous terrain at the edge of the African shield to reach the Congo basin was achieved by the colonial railway engineers only by using metre gauge, and tight turns and steep gradients. The capacity of the Chemin de Fer Congo-Ocean is therefore severely limited. So is the capacity of the Zaire railway, which also needs upgrading. That, however, is not the principal problem for Zaire's transport. Shaba province produces nearly all of Zaire's mineral exports. It is connected by a long and difficult railway route through Zambia, Zimbabwe and Mozambique to Beira. This route has been effectively shut because of the civil war in Mozambique, but could soon be rehabilitated. Second, it has a route through Tanzania to Dar-es-Salaam, but that involves transhipment across Lake Tanganyika. Third, it has a direct rail route which used to be the most heavily used across Angola to Lobito. Again, it seems at present as if peace might be restored in Angola and the rail link reliably restored. The only truly Zairean link is to Kinshasa – either by Kindu on the Lualaba or via Ilebo on the Kasai River; both involve transhipment and a lengthy river section. There are proposals to build a link between Ilebo and Kinshasa but for the moment they are just plans.

Agriculture and food supply

The most extensive area of agriculture in the region is to be found in the more humid south of Chad, which has proved excellent cotton-growing land. For much of the rest of Central Africa, agriculture is also the dominant employer (see Table 3.2). Congo has the lowest employment figure for agriculture at about 60 per cent (in earlier years quoted as low as 45 per cent): the rest are nearer 70 per cent or even higher. But agriculture accounts for only 13 per cent of the land surface overall. This is not just because of the low settlement density, it is also because there have been and still are other ways of collecting food from the forests – the pygmies of the Ituri basin of Zaire still operate a traditional gathering economy of collecting nuts, fruits, small animals, snails etc. Their impact on the forest ecosystem at low densities is negligible. Low levels of agricultural development are also because the high temperatures and high rainfall mean that for much of the area, even including the higher and cooler hills of Burundi and Rwanda, soils have a low organic content, are poor, and easily degraded. Many of the agriculturalists of the drier margins of the forest use slash and burn methods, growing on a plot for not more than a few years, then leaving secondary forest to regenerate for a fallow of fifteen or so years. (Most traditional systems are careful to avoid damaging large trees, useful for shade, and trees with particular fruiting value.) The sites of some of these people become more fixed if they are also involved with commercial crops such as cocoa. That too ultimately reduces soil fertility, but the trees may bear fruit for thirty or forty years before a fallow of equal length is necessary. This means that although the actual cultivated area at any one time may be small, the capacity for agriculture to degrade a much more extensive area is very high. To that extent, if the growing population numbers result in growing rural densities, then environmental damage is possible.

Central African countries have considerable difficulty in feeding themselves not just because of these environmental limitations. In addition government policies have in nearly every country put extreme pressure on agricultural prices. Overvaluation of the currency in some cases (the non CFA franc countries) means lower rewards for exporters, and commensurate possible gains for importers. State purchasing boards where they have been involved have also kept prices low in the interests of the urban rulers and attempts to keep the urban masses quiescent. The results have been distorted situations as in Congo, where in 1989, 40 per cent of its foodstuffs were imported – much of it of course up the difficult railway line to Brazzaville. For a country whose external debt service:export ratio in 1989 was 38 per cent, this is crippling and inexcusable. A similar problem in Zaire has led an EC/Zaire team to negotiate an aid project to restimulate and rehabilitate commercial farming in the hinterland of Kinshasa. Having the single and major market of Kinshasa is no doubt of great advantage: however, the long-term sustainability of agriculture in these environments is, as discussed above, not certain. In those countries like CAR, Cameroon and Chad which have extensive savannahs, it might appear that there is scope for increasing animal husbandry, particularly in the drier areas. But the tsetse fly remains the guardian of most of the rangelands, ruling out any such endeavour for the foreseeable future.

The 'failure' of agriculture means that the average calorie level of the diet in these countries remains below that of India, despite the very real progress that has been made in some aspects of GDP generation. What remains unknown is the extent to which poverty and malnutrition is concentrated in particular susceptible groups, and why.

Industry

Policy has in most countries attempted to foster the development of industry. Energy is often in short supply, and many raw materials have to be imported. Yet in some limited sense the policy has succeeded, in that there are indeed new industries, but the internal markets remain small and poor, and many industries have turned out to be poorer and more expensive suppliers than the importers for which they have substituted. Manufacturing, much of it small scale, only contributes around 10 per cent of GDP – in Equatorial Guinea only 1 per cent. The years of attempting to find a way out of neo-colonial dependency have not really worked well, though some favoured individuals with access to the relevant licenses have prospered. One observer commented (Ochola 1972: 135).

> Africa has behaved like the biblical five foolish virgins who squandered their resources and then had to go begging. The western theoreticians have advanced a concept that so long as there is an increase in per capita income, there is supposedly some development. This notion has been embraced by the under-developed countries to the extent that they have increasingly devoted their attention to per capita income growth rather than to development. (These development plans) result in the lowering of standards of living of the general masses.

There is, always has been, an alternative. Throughout the first half of the 1970s and 1980s, it is believed that the informal sector (IS) gained prominence as an important income yielding channel. But research into the Central African informal sector is still scanty. The bulk of existing works on these countries' informal sector is either brief (for instance various aspects on Gabon's informal sector have frequently appeared in the national newspaper L'Union,) or has remained unpublished, in BEAC (Banque Equatoriale de L'Afrique Centrale) reports for example. It is therefore urgent that the informal sector be researched since existing figures suggest the importance of this sector in trade. In Gabon 50 per cent of trading activities in the capital are said to be carried out by the informal sector (IS). Increased interest in the informal sector could also have positive consequences on the cultural evolution of Central African countries, building on their arts and cultural artefacts. However, dealing with the informal sector is not something which comes naturally to most bureaucracies. Because the idea of 'informal' is akin to 'illegal' the IS is often thought to be synonomous with 'illegal dealings'. Just as it would not be strictly correct to assume that the formal sector (FS) should be linked only to what is legal or legitimate, the IS ought not to be thought of as necessarily outside of the realm of legal systems.

However, there is indeed an extensive 'underground' economy in most Central African states. Among these underground activities are for instance dealings in gold and diamonds, and an illegal trade recently revealed in Gabon in ivory for the Japanese market – although it now seems as if the trade has been finally eradicated, but clearly the trade does not publish statistics! Some of these activities are short-lived because they find their *raison d'être* in temporary situations which result from the sudden break-down of the formal sector. For example, following a series of riots that broke out in Brazzaville, Zaireans from Kinshasa indulged in smuggling petrol across the river.

One problem with the IS is that although it may generate revenues such as 'space-use' taxes, mostly it is not eay to tax, at least formally. Congo has one minister per 100,000 inhabitants, and 160,000 civil servants, one for every thirteen inhabitants (*The Courier* 1991: 29). Most of these are on low but stable salaries. They have two needs of the IS. One is to maintain its 'illicit' nature, so that it can be harrassed for protection handouts. The other is that they depend a great deal on the IS for the provision of goods and services as the IS prices are generally low, and indeed seem lower in all countries in the region with the exception of Gabon. Extensive projects have been carried out in Rwanda with the aim of promoting IS activities, particularly in the *secteur non structure moderne* (SNSM). A number of BEAC reports on Chad have recognized the importance of the IS, which became more important as the civil war disrupted the FS.

Energy

The Zaire River basin has the second largest basin area and second highest discharge in the world after the Amazon. The basin contains one-sixth of the world's hydro-electric power (HEP) potential – and many believe it not unrealistic to contemplate exporting power by cable to Europe some time in the twenty-first century. One-quarter of this potential is at the site of the Inga dam, downstream from Kinshasa, which with European Community help is in the first stages of power generation. The power is transmitted, expensively, to the mining province of Shaba. To do that, access roads for the construction of the lines have had to be built right across the basin. But, as happened in Uganda with the Owen Falls Dam, the settlement density beneath the line means that it is not economical to install step-down transformers and local distribution. No intervening area therefore is in receipt of any benefits. Energy also means oil, in which two states are well endowed. It does not mean coal – none of these countries has other than minimal coal industries. So without oil and HEP, Chad, Cameroon, CAR, Rwanda and Burundi are finding 'development' difficult, and the prospects for favourable change are small. The most grandiose schemes are currently being trailed in public. The project suggested by a Nigerian Engineer to build a Transaqua Canal from the Zaire River to replenish Lake Chad would also generate hydro-electricity at several points along its length, therefore in theory energizing new industrial townships in Zaire, CAR and Chad (Pearce 1991). (See next section also.)

ENVIRONMENT

There are two kinds of concerns over the environment – over the extent to which there may be greater climatic changes occurring which can upset local ecosystems and economies, and over the extent to which human activity locally is degrading the environment directly.

Clearly the climate in areas like north Chad has seen major changes within the Pleistocene. The Tibesti massif was much wetter only a few thousand years ago, and cave paintings of hippos and other aquatic life demonstrate the extent to which the climate has since dried out. For several years in the 1980s rainfall in the Lake Chad catchment area was below the long-term average, and the lake shrank to a reputed one-tenth of its previous size – so much so that irrigation schemes were literally left high and dry. The water intakes for one of Nigeria's most expensive schemes for the arid north-east of the country are now tens of kilometres from water (Pearce 1991) and local people have moved on to the lake bed to farm the sediments. This period coincided with political and military turmoil in Chad, which destabilized the normal pattern of grazing by nomads. The combined effects of all of these events led to concern over desertification. But in the years 1987–8 rainfall was high, and the impoverished pastures also began to revive. There was optimism that the Lake levels might also be restored, but they have not been. Now some engineers are proposing schemes on the scale of those once conceived in the USSR to reverse the Arctic flowing rivers into the Aral Sea. Here, the Nigerian proposal includes a canal running from the Zaire River to empty into Lake Chad via the River Chari. The politics of water are beginning to affect the water-rich in the same way that the politics of oil affect the oil-rich as well as the oil-poor.

For the rest of these territories, there seems to be little current concern over climatic change. Rainfall is on average reliable –

although in the unlikely event of truly massive deforestation that statement might have to be reviewed. There are suggestions, however, that in past dry periods the rain forest of Africa shrank back to small residual islands. There are only about 7,000 plant species in the most diverse ecotones in the forests – fewer than in Amazonia and Asia. There are also forest margins which have a greater number of deciduous trees, and these are the margins which are most likely to be fired for agriculture.

There is an argument that suggests that the best protection for these forests is commercial exploitation, because then concerned and sustainable forestry exploitation will seek to protect the trees. On this line of reasoning upgrading the railways to the interior and building new ones can help protect forests. The historical evidence does not point the same way. There are significant areas of over-exploitation. Most of these are near the coast, in areas which have for long had better accessibility. The EC is assisting in reforestation schemes in Bas-Zaire. At the other extreme, in Rwanda, there is very little tree cover left. Local farmers are aware of the problems of soil erosion, and of the scarcity of fuel. They have willingly joined in schemes to improve and extend terracing which reduces soil erosion, and in tree planting.

Direct pollution is an unknown quantity. The low density of population means that air pollution by fire and energy use is low, and although locally in large cities waste disposal is an acute problem, the long-term future for such problems is good. The major problems concern the exploitation of heavy metals in so many mining areas. These inevitably cause contamination, both in extraction and concentration. These areas are also mostly upstream in the catchments. Little has been written about the problem, but equivalent exploitation in Amazonia has had widespread and devastating effects, particularly on water quality and through that the whole ecosystem.

One environmental resource problem has received massive media attention, namely the fate of the African elephants. All these countries had substantial elephant herds twenty-five years ago, and in nearly all cases the herds have now almost disappeared. In the case of the Central African Republic, their one-time self-crowned Emperor, Jean Bokassa, was almost certainly personally implicated in the slaughter and exploitation. But these are not the only animals to be at risk. Other species may not be hunted so much for their economic value, as for their local food value. This includes various small monkeys, civet-cats and even leopards.

EXTERNAL LINKS

The external linkages of these economies at independence were quite clear. Each country had well-established export and import links with the relevant metropolitan power – be it France, Spain or Belgium. The significance of these links varied greatly: Burundi and Rwanda had little external trade in relation to their low GNP, and were (they were at the time administered as a single colony) governed as a self-sufficient rural community. They are still (see Table 3.5) in this category. At the other extreme the development of the mines in Shaba (Katanga) lead to a major import-export trade. These two areas were also different in another way: the degree to which, low in the former, high in the latter, there were expatriate settlers. Europeans were also often in charge of the export crop plantations, such as for cocoa and coffee. The political instability at independence caused a massive efflux of expatriate populations, and with them the contacts and skills which had been inadequately engendered in the colonized peoples. For this reason as much as for the corrupt incompetence of the new administration, in a country like Equatorial Guinea the principal export, cocoa, dried up. In Katanga, the performance of the mines deteriorated significantly for a number of years (also of course affected by the Angolan civil war). Given the problems that the new states encountered, new arrangements were possible for those with the right ideological credentials, with Eastern Bloc countries. Gabon and Congo in particular were able to develop new barter trade links. In one sense successful, these links themselves have now been broken by the events in Eastern Europe. So there is again another reorientation. There is no doubt now that Europe continues to represent the major external partner for all these countries. It is also the case that the former colonial power, France or Belgium, is clearly re-established as the major partner. In a sense therefore 'Désir de changement et status quo semblent aller de pair': except that expatriates have not returned as private owners of crop plantations, and except that for Gabon and Congo in particular they now have a massive dependence on a new export wealth – oil. It is the major mineral exporters, these two countries plus Zaire, that are in export surplus.

The trade pattern emphasizes Europe. National leaders are obviously well aware of this, and they would much prefer to have closer internal regional ties. The development of political and cultural associations is important, as is the creation of new institutions like

Table 3.5 Principal exports and value of exports and imports in Central Africa

	value US$m imports 1988	exports	exports as % imports	US$ value of exports per capita	principal exports
Burundi	211	84	40	15	food + agriculture 82%
Cameroon	1,749	829	47	71	food + agriculture 55%
CAR	87	85	98	30	
Chad	181	131	72	23	food + agriculture 83%
Congo	580	1,087	187	467	fuel, energy 93%
Equatorial Guinea	25	23	92	66	
Gabon	686	1,475	215	1,161	fuel, energy 80%
Rwanda	296	131	44	18	food + agriculture 90%
Zaire	794	1,003	126	29	minerals, ores 71%

Source: Encyclopaedia Britannica (1990).
Note: Dominant partner for all exports and imports EC in all cases.

the Bantu CICiBA (mentioned above) important. The OAU-sponsored Kano conferences were important in resolving Chad's problems. However, none of these agencies yet has the economic clout of those stemming from the North.

THE GEOGRAPHY OF IGNORANCE

Patterns of power and dependency are never static, but always open for renegotiation, and induced change. In some parts of the rain forests, there is a conception of how the pygmy Twa peoples were in equilibrium with their environment (Pagezy 1985), and also in trade relationships with the few agriculturalists at the margin of the forests. Whether or not this ever did represent a 'true' picture of some undated past, we know that changes in relationships are in practice continual. The Twa are becoming more closely tied, often as labourers, to farmers and their plots. They have not given up their hunting, but they are more likely to hunt only in restricted ranges. Is this why the game is disappearing, and if so is it only in some locations? Are other locations relatively less molested than before? Now that the Twa become more fully incorporated in a wider economy, are they an underprivileged sub-class? How do these social changes preferentially affect the diets and health of women and men, old people and young children? Are rural people as a whole an underclass for the new urbanites? How do the new urban classes view their responsibilities to the rural areas and these globally significant forests? What education is relevant to Africa? Are the educated unemployed a resource or a threat? What attitudes prevail within the institutions of government and education? Is it a good idea environmentally to change these societies into food-importing urban polities? If food is to be produced locally, how will agriculture accommodate simultaneously the three 'Ss' – sedentarism, sustainability and surplus? On nearly all these issues the answer is, we do not know, nor is there any likelihood that we will know for a long time yet. There are too many powerful politicians who prohibit social science research, and in any event the education budget is still too small in most of these states to sponsor such quests for apparently esoteric knowledge.

The problem is that the question posed at the beginning of this chapter, that at independence there was no model of development for these states to follow, has not been answered by the changes of the last twenty-five years. Nor will the question be answered, until such time as all these 'lesser' questions are placed on the research agenda.

REFERENCES

Aicardi De saint Paul, M. (1989) *Gabon: The Development of a Nation*, trans. A.F. Palmer and T. Palmer, London: Routledge.

Allen, C., Radu, M.S., Somerville, K., and Baxter, J. (1989) *Benin, The Congo, Burkina Faso*, London: Pinter.

BEAC (1989) *Sommaire Statistiques Economiques* 163, Banque Equatoriale de L'Afrique Centrale.

Bezy, F. (1990) 'L'évolution economique et sociale du Rwanda depuis l'Indépendence', *Mondes en développement* 69 (18).

The Courier (1982a) 'Rwanda: the shadow of Malthus over a genuine hope of economic progress', 72: 25–31.

The Courier (1982b) 'The Congolese Forest: a vast resource hardly exploited', 74: 58–9.

The Courier (1983) 'Zaire', 78: 26–54.

The Courier (1984) 'Congo: country report', 83: 22–39.

The Courier (1987) 'Rwanda: a thousand hills, a handful of options', 105: 12–30.

The Courier (1988a) 'Equatorial Guinea: what sort of development policy?', 107: 32–42.

The Courier (1988b) 'Zaire: the interminable restructuring', 110: 12–31.

The Courier (1991) 'Congo: a delicate transition to democracy', 125: 27–45.

Cruise O'Brien, D.B., Dunn, J. and Rathbone, R. (eds) (1989) *Contemporary West African States*, Cambridge: Cambridge University Press.

Degand, J. (1990) 'De l'autosuffisance a l'économie d'échanges en agriculture: Le cas du Burundi', *Mondes en développement* 18: 49.

Dumont, R. (1962) *L'Afrique noire est mal partie*, Paris: Seuil.

Encyclopaedia Britannica (1990) *Book of the Year*, Chicago.

Geographical Digest (1964)

Geographical Digest (1966)

Geographical Digest (1970)

Geographical Digest (1980)

Harrison, J. (1970) 'Industry and crafts in Madagascar', *La Croissance urbaine en Afrique Noire et Madagascar (The Urban Growth of Black Africa and Madagascar)*, vol. 1, Centre National de la recherche scientifique, London: Bogles l'Ouverture.

Jeune Afrique (1990) 'Gabon: Bongo condamne au multipartisme', 1529, 23 April.

Lassère, G. (1970) 'Les méchanismes de la croissance et les structures démographique de Libreville 1953–1970' unpublished paper.

Lavroff, D.G. (1970) *Les Partis politiques en Afrique Noire: Que sais je?*, Paris: Presse Universitaire Française.

Mba, J.F. (1987) *La Recherche en science sociale au Gabon*, Paris: Harmattan.

Meister, A. (1966) *L'Afrique: peut-elle partir?*, Paris: Seuil.

Ministère de l'Information, Government of Gabon (1965) *5 ans de Gabon* (official publication).

Mountjoy, A.B. and Hilling, D. (1988) *Africa: Geography and Development*, London: Hutchinson.

Ochola, S.A. (1975) *Minerals in African Underdevelopment*, Paris: Bogles l'Ouverture.

O'Connor, A. (1991) *Poverty in Africa: A Geographical Approach*, London: Bellhaven Press.

Pagezy, H. (1985) 'The food system of the Ntomba of Lake Tumba, Zaire', in Pottier, J. (ed.) (1985) *Food Systems in Central and Southern Africa*, London: School of Oriental and African Studies.

Pearce, F. (1991) 'Africa at a Watershed', *New Scientist* 1761: 34–40.

Peemans, J.P. (1990) 'Le Burundi du IIIeme au Veme plan: Contraintes de modernisation et enjeux de développement', *Mondes en développement* 18: 39.

Pottier, J. (ed.) (1985) *Food Systems in Central and Southern Africa*, London: School of Oriental and African Studies.

Thompson, V. and Adolff, R. (1960) *The Emerging States of Equatorial Africa*, Oxford: Oxford University Press.

Wade, A. (1990) 'Pour les partis panafricains', *Jeune Afrique* 1556 (28 Oct.): 24–30.

Wautelet, J.M. (1990) 'Cameroon, accumulation et développement', *Mondes en développement* 18: 75.

World Bank (1984) *Toward Sustained Development in Sub Saharan Africa: A Joint Program of Action*, Washington DC: World Bank.

World Bank (1990) *World Development Report*, London: Oxford University Press.

World Today: Africa (1989)

Ziegler, J. (1980) *Main base sur l'Afrique: Recolonisation*, Paris: Seuil.

4

THE CHANGING GEOGRAPHY OF WEST AFRICA

Kathleen Baker

IMAGES OF WEST AFRICA, PAST AND PRESENT

When The Gold Coast became the independent state of Ghana in 1957 and the first independent state in West Africa euphoria swept the region, stimulating in turn greater pressure for independence by other states. By 1965 all former French and British territories were independent as was Liberia, unique because it had never truly been a colony (see Figure 4.1). The one exception was Portuguese Guinea, now Guinea-Bissau, which was involved in a long and destructive struggle with the Portuguese, but which finally won its independence in 1974. Thus from a single region dominated by the colonial administrations of Britain and France, and to a far lesser extent by Portugal, there emerged fifteen independent states, each with leaders with very different, but clearly defined political ideologies – some Marxist, such as Benin, and some with strong capitalist orientations such as the Côte d'Ivoire (Ivory Coast) – on how social and economic aspirations could be realized in a region where living standards of the majority were still painfully low. What was particularly interesting was the attitudes of the new states to decolonization. At one extreme, Kwame Nkrumah in the guise of a national liberator chose to sever ties with Ghana's colonial past, while at the other extreme, Houphouët Boigny of Côte d'Ivoire retained very close links with France as according to Dunn (1978: 12–13) he was either 'a lackey of international capitalism, or a paragon of pragmatic rationality'. Attitudes of other states have been eclipsed to some extent, by the new leaders of Ghana and Côte

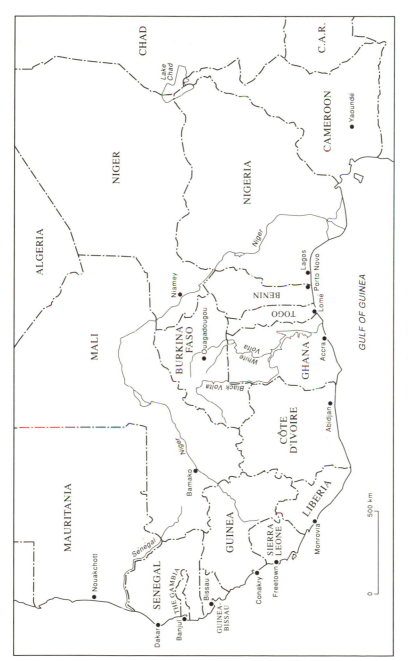

Figure 4.1 West Africa

d'Ivoire, but there were clear divisions. Senegal chose to retain ties with France, though was not as close as Côte d'Ivoire, while Mali and Mauritania chose to move away, rejecting even the common currency of the francophone states, though Mali has now returned to the franc zone. And Guinea, which had antagonized President de Gaulle of France by asking for total independence a few short months before France was ready to grant it, was abandoned by Paris and forced to go its own way. But for all the choice that freedom implied, the reality of independence has been much more tightly constrained.

Poverty and low living standards are as widespread now as they were at independence, harsh evidence that there has been little improvement in social and economic conditions over the past twenty-five to thirty years. As further evidence of the region's decline and stagnation, thirteen of the fifteen states of the region are now classified by the World Bank as being among the world's poorest. What is particularly significant about this is that efforts to develop the region have consumed vast resources since independence, but success has been minimal. Many errors – more easily identifiable with hindsight – have been made in the development struggle, and what does seem clear now is that a radical change of approach is needed, a change which incorporates Africans to a very significant extent, in the search for solutions to African problems.

POPULATION AND PEOPLES

The population totals of West Africa's fifteen nations are still essentially unknown: the *estimated* population of the region is thought to be low at around 200 million. With estimated population growth levels of around 3.0 per cent during much of the past two decades (World Bank 1989: Table 28), the region's population is set to double in around forty years. However, it is now believed that population growth is either slowing down or never was as great as had been forecast because total populations had been wrongly estimated.

One of the biggest question marks hangs over the size of Nigeria's population, believed to account for half the people in West Africa. Corruption of the results of Nigeria's 1973 census has resulted in the questionable 1963 census being used as a base line (Kirk-Greene and Rimmer 1981). Extrapolation of population data over thirty years must conceal gross inaccuracies, and in Burkina Faso – formerly Upper Volta – population projections over a similar period were some 12 per cent below the 7.92 million people counted in the 1985 census (Economist Intelligence Unit 1986–7). If similar discrepancies are true for other states as well, then the population of the region could be far higher than current best estimates suggest.

While fertility levels have remained fairly constant at six to seven live children per woman, and the crude birth rate too has remained stable with forty-five to fifty live births per thousand people, an important contributory factor to increasing population is improved health care. Life expectancy has increased and is now nearer fifty years than forty years as it was in the mid-1960s. Death rates too have fallen with several nations showing declines of between 30 and 50 per cent over the past twenty-five years. In spite of achievements in health, these must be seen in relation to conditions in the West. Life expectancy is some twenty years less in West Africa than in Western Europe, and even the substantial reduction in death rates to around eighteen per thousand in West Africa still far exceeds equivalent figures of twelve and ten in the UK and in France respectively.

One of the most encouraging statistics is the decline in infant mortality levels in every state and by as much as 65 per cent in Nigeria (World Bank 1989: Table 31). At the same time alternative sources cited in Dixon et al. (1989) reveal that infant mortality levels in Nigeria's urban areas are extremely high, being positively correlated with overcrowding and other variables.

Without reliable national statistics the validity of any data is in question. But whatever the discrepancies, all statistics show that in spite of reductions in infant mortality of around 40 per cent in the region over the past twenty-five years, the levels are still unacceptably high. They range from 171 deaths per thousand live births in Mali, to Ghana and Liberia with the lowest levels of 92 and 89 respectively, but the Liberian figures will increase again as a result of the effects of the brutal civil war of 1990. Even the lowest rates in West Africa are unacceptably high when compared with Europe, where in the UK and France infant mortality rates have fallen to nine and eight per thousand live births respectively. Equally depressing are statistics for the mortality levels of the under fives. These too far exceed levels in Western Europe and are noticeably higher in the northern states of West Africa where they exceed 150 deaths per thousand live births. In the southern states the figure is little better at 100–150 deaths per thousand live births.

Throughout the region people are aware of improving health

facilities. The ratio of doctors and nurses per head of population has improved over the past twenty-five years but on average it is still about 1:14,000 people over the region as a whole; while comparable ratios in the UK and France were 1:680 and 1:460 respectively in 1987 (World Bank 1989: Table 33). Immunization programmes have made major improvements to health and the incidence of diseases such as malaria, measles, polio, diphtheria and cholera is decreasing, a fundamental prerequisite for development (Timberlake 1985). AIDS presents a rapidly growing threat though WHO statistics (1990) indicate that there are fewer cases in West Africa than in other parts of the continent. While AIDS could dramatically reverse the current trend of population growth, in the foreseeable future, it seems fairly certain that even if rates of increase have slowed, a period of sustained growth in West Africa's population is inevitable.

NATIONS CAUGHT IN ECONOMIC TRAPS

Agriculture dominated the economies of West African nations at independence, the greater part of GDP being derived from low value, relatively unprocessed primary produce. Some three decades later agriculture may seem less important in relative terms (Table 4.1), but in absolute terms the numbers involved are as great (World Bank 1989: Table 34). There has been growth in industry in recent decades, mainly due to the development of mineral resources – oil in Nigeria and uranium in Niger for example – but the main change is that the service sector is now very often the chief contributor to GDP rather than agriculture.

West African economies have been characterized by limited growth and in many cases stagnation throughout much of the independence era. Nations such as Togo, Côte d'Ivoire and Nigeria showed positive rates of growth in the 1965–80 period, largely due to benefits derived from commodities such as phosphates, coffee, cocoa and petroleum. These growth levels have not been maintained and economies have since grown more slowly. Others such as Burkina have shown higher growth rates more recently, but overall growth rates have been lower in the 1980–8 period throughout much of the region (Table 4.2). Data showing GNP on a per capita basis merely confirm the trend of decline and stagnation as ten out of fifteen West African nations have had negative growth rates in the 1980–7 period.

Table 4.1 Structure of production in West Africa

| | Distribution of gross domestic product (per cent) | | | | | | | |
| | Agriculture | | Industry | | Manufacturing[a] | | Services | |
	'65	'88	'65	'88	'65	'88	'65	'88
Burkina Faso	53	39	20	23	—[b]	13	27	38
Mali	65	49	9	12	5	5	25	39
Nigeria	54	34	13	36	6	18	33	29
Niger	68	36	3	23	2	9	29	41
Togo	45	34	21	21	10	8	34	45
Benin	59	40	8	13	—	6	33	47
Ghana	44	49	19	16	10	10	38	34
Guinea	—	30	—	32	—	5	—	38
Mauritania	32	38	36	21	4	—	32	41
Liberia	27	37	40	28	3	5	34	35
Sierra Leone	34	46	28	12	6	3	38	42
Senegal	25	22	18	29	14	19	56	49
Côte d'Ivoire	47	36	19	25	11	16	33	39

Source: World Bank (1990: Table 3).
Notes: [a] Because manufacturing is generally the most dynamic part of the industrial
sector, its share of GDP is shown separately (World Bank).
[b] Data unavailable.
Data unavailable for Guinea-Bissau and The Gambia.

Table 4.2 Growth of production in West Africa

| | Average annual growth rate (per cent) | |
	1965–80	1980–88
Burkina Faso	—[a]	5.5
Mali	3.9	3.2
Nigeria	6.9	−1.1
Niger	0.3	−1.2
Togo	4.5	0.5
Benin	2.1	2.4
Ghana	1.4	2.1
Guinea	—	—
Mauritania	2.0	1.6
Liberia	3.3	−1.3
Sierra Leone	2.8	0.2
Senegal	2.0	3.3
Côte d'Ivoire	6.8	2.2

Source: World Bank (1990: Table 2).
Notes: [a] Data unavailable.
Data unavailable for Guinea-Bissau and The Gambia.

Principal causes of poor economic performance

Both internal and external factors have contributed to poor economic performance in West Africa. The size and distribution of states has done much to hinder development. On the one hand states such as The Gambia and Guinea-Bissau are so small and resource poor as to be economically unviable, while on the other hand, some of the Saharan states are so large as to be effectively isolated from themselves and the rest of the region. The boundaries could have been drawn better. The nature of the inherited economies dependent on exports of low value, relatively unprocessed primary produce is yet another factor that has left West African nations ill-prepared to adjust to rapid changes in the global economy. Failure of post-independence governments to invest adequately in agriculture, the main foreign exchange earner, resulted in reduced output for export and loss of foreign exchange. This has been exacerbated by droughts in the early 1970s, and again in 1983–4. Low value exports have been paralleled by imports of processed products and fuels, all of which have risen rapidly in value, and have contributed significantly to the region's balance of payments problems. Furthermore, the population of the region has been, and still is, relatively small in relation to its geographical size, and this has constrained the development of markets, though rapid population growth could see this change.

External factors have added significantly to economic problems: fluctuating prices and demand for West African products have partly reflected global economic conditions, and partly the development of substitutes by former importer nations. For example, the array of oil seeds available in the West has reduced demand for palm and groundnut oils, and synthetic compounds have increased alternatives to rubber. The oil shocks of the 1970s put pressure on the balance of payments of most West African nations, though they did brighten Nigeria's prospects. The region's coffee and cocoa producers benefited from the beverage boom, and backed by increased revenue from these commodities, governments borrowed heavily, investing in projects many of which proved uneconomic or unsustainable in the long term. In the aftermath of the boom, West African nations found themselves trapped in debt. Indebtedness has drained West Africa and has been made all the worse by high interest rates and by the strength of the dollar, the currency in which the majority of loans are quoted. Although West Africa's total debt of some $57 million in 1987 was not large in itself, it is still a major burden on individual nations. The debt service alone is equivalent to between 15 and 30 per cent of the value of goods and services exported by most states and, in the case of the Côte d'Ivoire, is as much as 40 per cent. An additional factor in Côte d'Ivoire, and also it would seem in other states, was that increased revenue during commodity boom years led to the expansion of patronage and with it the expansion of parastatals, noted for their corruption and inefficiency, all in the name of development.

Overvalued currencies

Many of West Africa's economic problems result from budget deficits caused by governments spending far in excess of their revenue. With little scope for covering these deficits through public borrowing, taxation or other non-inflationary means, most governments have printed money to resolve the situation. The resulting inflation has led to currencies being overvalued, that is, where the official rate of exchange at national banks between the Nigerian naira, or the Ghanaian cedi, for example, and the US dollar, is maintained. For example the cedi was for long officially valued at 2 cedis to the US dollar, when it should have been nearer 30. This makes imports look very cheap, and generates an excess demand for dollars, which then have to be rationed to licensed and therefore privileged persons. It also means that exporters get paid very few cedis for their exports. Effectively they subsidize the favoured importers. The overall imbalance depresses exports, favours selected importers, and generates smuggling and black market corruption.

In the case of Ghana, cocoa has long been the main generator of foreign exchange. According to Chazan (1983), net payments to farmers declined from 27.9 per cent of the real value of cocoa production in 1973–4, to 8.08 per cent in 1978. Farmers were thus getting so little from cocoa that many farms were abandoned or the cocoa was sold on alternative markets (see below). Overall production in Ghana was recorded as being around 19 per cent lower in 1977–8 than in 1973–4, and as a result cocoa's contribution to GDP fell by 5 per cent per year on average from 1973 to 1978.

The rural areas were further penalized because between 1973 and 1978 retail food prices rose by 450 per cent and consumer prices rose by 200 per cent. While receiving low farm prices for their cash crops, the rural areas were forced to pay inflated prices for consumer goods which left progressively less for reinvestment in agriculture, so hastening its deterioration. With prices so unrealistically high and the availability of goods so limited, the standard of living of Ghanaians in both rural and urban areas declined. Methods of surviving the crisis were varied and revealed the initiative of Ghanaians. Many simply migrated to other parts of Africa, but in Ghana a parallel or black market economy soon flourished – the *kalabule* economy. A major part of this involved getting access to vital foreign exchange and smuggling was one of the most important ways of doing just this. Ghanaian products were smuggled across the border to Upper Volta (now Burkina Faso), Côte d'Ivoire and Togo where they could be exchanged for CFAs, the convertible currency of neighbouring francophone states. Smuggling was no small enterprise but a highly organized form of trade. For the estimated amount of cocoa to have been smuggled out of Ghana, Chazan (1983) cites estimates that 150,000 men would have had to have worked 150 days per year. Other commodities were smuggled out of the country as well, including 12 to 15 per cent of Ghana's gold and diamonds during the decade.

Donations of food aid, such as grain and powdered milk, were found on the open market at astronomic prices revealing the extent of contacts with those operating official markets. The many diversions of funds destined for the government coffers severely reduced state revenues and resulted in the economy sliding even further into chaos.

Nations with sovereign currencies, mainly the anglophone states and some of the francophone, could have devalued, but in nations such as Ghana and Nigeria for example, this option was long rejected as being politically unacceptable. Imports, for which demand was greatest in the urban areas, would have escalated in price with devaluation, jeopardizing the future of governments in power. Under its structural adjustment programme (see p. 90) Ghana has been induced to act courageously, and has devalued its currency, and Nigeria has achieved much the same effect by auctioning the naira since 1987. However, the option of devaluation is not open equally to all nations in West Africa. The majority of nations of former French West Africa share the same currency, the CFA franc, which has been tied to the French franc at the rate of FF 1 = CFA 50 since 1948. After independence Mauritania withdrew from the franc zone, adopting its own currency instead, and Guinea was left little alternative but to do the same. Mali too left the franc zone for a temporary period. Other nations have also shown interest in the CFA: Guinea-Bissau attempted negotiations to join in 1988, but is currently considering the possibility of union with other Lusophone states in Africa. More recently Ghana applied to join the franc zone but was rejected (Collier 1990).

The convertibility of the CFA has been one of its major attractions but recent research has shown that while franc zone countries fared better in economic terms in the 1970s, the reverse has been true in the 1980s (Lane 1990). In spite of positive feelings towards the CFA franc, the link has proved expensive for France, particularly in recent years as Africa's debts have mounted. The possibility of a devaluation of the CFA is very real, and while there are still no clear directives, what does seem possible is that the CFA franc may, in the future be realigned with the ecu, the EC unit of value based on a basket of European currencies, rather than the French franc (Collier 1990; Lane 1990). Devaluation of the CFA by one means or another, should fit in with Structural Adjustment in that it should promote exports and reduce demand for more expensive imports.

Structural adjustment: a cure for all ills?

One major change in the economic direction of West Africa in recent years has been the increased involvement of the IMF and the World Bank in response to growing indebtedness. In order to secure continued funding, the IMF together with the World Bank have pressured many states throughout Africa to adopt World Bank inspired economic recovery programmes. Prescribing more or less the same medicine for all, the aim of Structural Adjustment Programmes (SAPs) is to persuade governments to involve themselves less in African economies, leaving market forces to operate more freely. Measures vary between states, but austerity is everywhere the key word. The use of domestic produce as opposed to imports is being encouraged, and in order to stimulate domestic production in the rural areas, farm prices are being increased. Imports are being restricted where possible, and exports promoted in order to obtain essential foreign exchange. An important component of the austerity measures in many West African countries has been severe public sector salary cuts and redundancies.

Governments have had to tread warily as introducing salary and job cuts at the same time as food prices were rising, due to increases in farm prices, has been far from popular. Hardship is undoubtedly occurring and, without evidence of rewards in the form of improved living standards, has contributed to political instability. In Benin, the socialist government has been ousted, its unpopularity due to the enforcement of measures designed to adjust the country's economic structure. In Burkina Faso under Thomas Sankara (1983–7), a structural adjustment programme modelled on that of the IMF created considerable discontent in urban areas and arguably led to Sankara's downfall. In Nigeria where austerity measures have exacerbated urban poverty and malnutrition, there have been attempts to oust Babangida, but promises of return to civilian rule have done much to restrain those who would be rid of him. Ghana, by dint of Rawlings' will, has managed to avoid political instability but this remains a threat. Opinion is still divided as to whether or not structural adjustment is working in Africa. The World Bank is now aware of the human suffering that SAPs can provoke, and there are hopes that in the future structural adjustment will have a more 'human face'. The most recent review of these programmes is to be found in Harrigan et al. (1991).

Trade patterns have retained some features of the early days of independence though both imports and exports have fallen in value between 1980 and 1987. With some caution this may be explained as one of the results of SAPs but such conclusions are risky, and it is necessary to see whether the trend continues. Over 50 per cent of the value of West Africa's exports are still derived from primary produce and the second most important earners of foreign exchange are fuels and minerals. Export of industrial products is minimal and machinery and transport equipment, together with other manufactured goods make up about 65 per cent of the value of imports compared with about 75 per cent in 1965. Foods and fuels are the other main imports accounting for around 19 per cent and 12 per cent of imports respectively.

ECONOMIC ENTRAPMENT, GOVERNMENT INEXPERIENCE AND POLITICAL TURMOIL

West Africa has a bad image in terms of political instability, yet in one respect has a remarkable record. The region has seen fewer wars than most of the rest of Africa, and there has been less interference in the affairs of neighbouring states. The Biafra war for the secession of Eastern Nigeria was a notable exception, as was Mauritania's reluctant involvement in the Western Sahara, the war in Liberia and the brief conflicts between Mali and Upper Volta/Burkina, regarding a common boundary. One could argue that boundaries drawn by the colonial powers for their own convenience, many of which have divided ethnic groups such as the Ewe on the Ghana–Togo boundary, and the Wolof in the Senegambia, have served to keep the peace between states as members of the same ethnic group are reluctant to fight relatives on the other side of a totally artificial divide. This was indeed apparent in the Mali–Burkina war of December–January 1985–6.

West Africa has not been an easy region to govern mostly because of intra-state problems. By the mid-1960s, military governments had assumed power in a great many states. Dissatisfaction with Kwame Nkrumah led to his replacement by a military regime in 1966, and the government in Nigeria was similarly deposed in 1966. Maurice Yameogo was replaced by General Laminzana in Upper Volta in 1966 – the first of many coups – and Modibo Keita was replaced by General Traoré in Mali in 1968, who in turn was replaced by Amadou Toumani Touré in March 1991. There were also military coups in Togo (1963 and 1967), Benin (1972), Mauritania (1978) and Sierra Leone (1978). Togo experienced coups in 1963 and 1967 and, after a brief attempt recently to restore civilian rule, the army under General Eyadema has once more asserted its power. Indeed, change of leadership was fashionable, and for many states the fashion continued. While military regimes still dominate the region, there are a few notable exceptions such as Côte d'Ivoire, Senegal and The Gambia where civilian governments are allegedly democratic. What lies ahead for Côte d'Ivoire when the leadership eventually does change is unpredictable though it could have a destabilizing impact on the region, particularly if it involves withdrawal of the French and foreign capital. Both Senegal and The Gambia are anxious to ensure that democracy continues.

Explanations for political turmoil are numerous and may revolve around religious differences, conflict between richer and poorer elements of society, corruption and many other factors. However, two common problems which have faced governments of the region are considered here: first, ethnic rivalries, and second, the failure of governments to deliver the economic and social improvements promised at independence. West Africa's national boundaries frequently group together rival ethnic groups. At independence the unifying domination of colonial governments was instantly replaced by rivalry between ethnic groups. In Nigeria, for example, reconciling the interests of the three major divisions: the Muslim North, the Yoruba West and the Ibo East has been a nightmare, deflecting the attention of successive Nigerian governments from the more usual tasks of government. The same is true for other states: Mauritania, for example, has suffered from rivalry between northern Arab peoples and southern negroes and the brutal Liberian civil war of 1990 was the result of internal division aggravated by Libyan interference in the early stages. In spite of this, ethnic friction is far greater in other parts of the continent.

While the British colonial governments went some way to preparing African nations for independence – albeit on the Westminster model – little was done to promote feelings of nationalism. One of the reasons for Ghana's instant 'success' at independence was that Nkrumah created nationalist feeling by adopting the name of a former ancient West African empire – Ghana – for the new state, a name of historic importance with which all ethnic groups could identify. In spite of this, Ghana was soon in no better a position than Nigeria: regional disparities persisted and ethnic rivalries flourished. Although national identity does exist, ethnic affinities are often still more important than sentiments about the nation.

Failure to bring about improved social and economic conditions has also contributed to instability. While many problems facing governments have been beyond their control, there has also been evidence of bad decision-making. Neglect of the rural sector, over-investment in non-productive schemes and the persistence of overvalued currencies in many states all testify to errors in decision-making. Since the early 1980s there seems to have been greater awareness that replacement of inexperienced military leaders with more of the same was solving no problems. The frequency of coups has diminished, but several factors could provoke further political turmoil, not least the impact of SAPs (see pp. 89–90). There are, none the less, high hopes that several nations will exchange military rule for multi-party democracies though the recent experience of Togo is not encouraging. Elaborate rules have been laid down for the establishment of a multi-party democracy in Nigeria, but there is concern that change from military to civilian rule may mean a little more than the relaxation of some regulations, and a change of clothes for leading personnel in the military government.

A major change is that the political ideologies of governments, clear at independence, appear to have weakened and virtually disappeared as nations have been overtaken by economic and political strife. And now, under the direction of the IMF and the World Bank, political ideology seems to have been eclipsed altogether in some states or replaced by the pragmatism necessary for implementing SAPs. It awaits to be seen whether readjustment of IMF strategies to give them more human and ecologically sound foundations will also result in greater latitude for policy implementation of individual governments.

URBAN EXPANSION: POPULATION INCREMENTS UNMATCHED BY RESOURCES

Urban expansion is one of the most notable changes in the geography of West Africa over the past twenty-five years. Only 7 per cent of West Africa's population was urbanized in 1965, and by 1985 the figure had reached 31 per cent. The relative growth of African cities has been unparalleled on a world scale.

Why towns and cities have grown

Rural–urban migration was the main cause of the rapid urban growth that West Africa experienced in the 1960s and 1970s, though evidence in the 1980s suggests that the rate is slowing down. Migration, however, is still the main cause of West Africa's increasing urbanization. People move to cities for a variety of reasons most of which are usually bound up with the expectation of a better quality of life. Higher incomes combined with a less poverty-ridden existence than in the rural areas, access to better health facilities, better schooling, the sheer desire to escape from family pressure in the village, and to experience the 'bright lights' of the city are all common reasons (O'Connor 1983). Some twenty to thirty years ago it was generally males who were associated with migration, but increasingly, whole families are moving to the city. The type of migration varies from city to city. Dakar, for instance, has long been the recipient of migrant families, whereas in the early days of oil, migrants to Lagos were generally young men whose families either stayed in villages or joined them later. Job opportunities are perhaps fewer now in many urban areas than they were in the early 1980s, due largely to economic decline and the imposition of austerity measures; for women there have always been fewer job opportunities than for men in urban areas.

The droughts, particularly in 1973 and again in the early 1980s, also forced people to move to urban areas. The northern states suffered far more than did the south, and loss of livestock and crops resulted in the movement of thousands to towns. Nouakchott in Mauritania expanded from a few thousand in the early 1960s to a city of about half a million people now. Urban areas in Senegal, Burkina, Mali and Niger also expanded with each successive period of drought, and the effects extended further as people from the north moved south in order to find work.

The commodity boom of the early 1970s which included the first major oil price increase also did much to promote urban growth in West Africa, particularly in the region's southern cities. The growth of industry mainly in Nigeria, and the service sector everywhere, increased employment potential for many urban migrants. In Côte d'Ivoire, for example, revenue from the beverage boom was used for investment in a vast range of projects the administrative centres for which were based in urban areas and provided increased employment opportunities.

While migration was responsible for the rapid urban growth of the 1960s and 1970s, natural increase is now the main contributor to continued urban growth, adding substantially to overcrowding and to pressure on already overused resources. Although the numbers classified as urbanized increases substantially each year, the growth rate of the urban population has decreased from 12.4 per cent in the 1960–85 period, to 3.4 per cent between 1980 and 1985 (World Bank 1989).

Some effects of rapid urban growth

Rapid growth of cities has overtaken planning, and prevented the implementation of quality standards. Development has been piecemeal. Schools and health facilities which in theory are more readily accessible than in the villages are nevertheless far from sufficient to meet the demands of rapidly growing urban populations. Water supplies are poor everywhere; provision for refuse collection is quite inadequate in most cities, and disposal of waste is even more difficult. In Dakar, for instance, several refuse tips formerly beyond the urban area are now completely surrounded by housing (Ngom 1989). Municipal budgets have been hit by falling tax revenue due to job losses and wage cuts. Many cities now depend on aid in order to improve drainage, sewage and water supply systems. Perhaps the biggest problem is housing because there is far from sufficient of it and where it does exist standards are pitifully low – except in areas occupied by black elites together with a remnant white population.

The elite districts near the Corniche in Dakar are but a short distance from the 'African quarter', where the density of housing and people is high, and where the stench is sufficient indication of an inadequate sewage disposal system.

White (1989) also argues that rapid growth of cities is adding to environmental degradation. As villages which surround towns become incorporated in the urban sprawl, wells become overused and in parts of the Cap Vert peninsula in Senegal, many have become saline, a combination of overuse and years of drought. Another way in which urban growth may be contributing to environmental decline is through the demand for charcoal. Through this, White (1989) also estimates that urban dwellers are exerting greater pressure on forest resources than do rural dwellers. According to Dumont and Mottin (1980), for every tree planted in Senegal, fifty are cut to serve the needs of Dakar. It is urgent that pratical policies are adopted regarding environmental health, in particular sewage, water supplies and housing quality, but it will be many years before the provision of services is anything like adequate and able to reduce the overcrowding where in parts of Lagos some sixty people may live in the same room (Peil 1990).

Most Africans do not depend on one permanent job but on several forms of income-generating activity. Definition of the 'informal sector' where people derive a living from a wide range of activities is difficult and O'Connor (1983) has referred to it as the small-scale sector. Here employment may or may not be regular, and it can be very varied. It may involve long distance or local trading, porterage, transport, the making and selling of herbal medicines and charms, or virtually anything for which there is demand. Even those in formal employment add to their income by part-time involvement in the small-scale sector. While women have far fewer employment opportunities in the large-scale sector in urban areas, they are greatly involved in the small-scale sector as are children. The importance of the small-scale sector cannot be underestimated in Africa, as work by Awassi clearly shows (Awassi 1990). It has always existed, but its true importance has been realized only within the past fifteen to twenty years.

THE INDUSTRIAL SECTOR IN WEST AFRICA

Industry in West Africa, most of which is centred in the urban areas is still far less important to the region than either agriculture or services (Table 4.1). Its share of GDP is still around 20 per cent on average for the region and in nations where it is more important such as Togo, Senegal Liberia, Mauritania and Nigeria among others, it is mining rather than manufacturing which is the main contributor. Manufacturing industry is still relatively unimportant and with the exception of Senegal, Côte d'Ivoire and Nigeria, it contributes under 10 per cent to GDP in most states.

Many of the resources necessary for industry would at first sight appear to be available in the region: for example metallic and non-metallic ores are plentiful. The larger mineral deposits such as Ghana's gold, Sierra Leone's diamonds, Niger's uranium and Nigeria's oil have been identified and are being worked. But there is much more. However, exploitation is often not economically viable because of inadequate infrastructure, particularly in communications, inadequate domestic technical expertise, inadequate finance or an unattractive political record for overseas investors, and in addition deposits are often small and spread over a large area. In Burkina Faso for example, small gold deposits are known to be numerous, manganese is abundant in the north-east of the country on the borders with Niger, and there are also plentiful deposits of gypsum and limestone, all of which might at some stage be valuable to the development of domestic industry.

Besides minerals, West Africa does have other resources which could form the basis of industry. The region covers three major ecological zones, the sahel, the savannahs and the forest, all of which could supply both cultivated and uncultivated products to industry. Another major resource is sunshine which is an important element in the tourist industry, and, as Onyemelukwe (1984) argues, could be used as an important power source in the future.

A major problem for West African industry is the small size of most of the domestic markets, and in addition it is difficult to combine adjacent markets for political reasons and because of poor inter-state communications. The alternatives are obviously export markets, but any attempts by West Africa to further process produce before export meets stiff competition from established producers with larger home markets and scale economies, and often discriminatory trade barriers as well. Trade blocks such as the European Community have hindered Africa's efforts to prosper from trade as much as they have helped through aid. One could argue that industrial development needs to be much 'lower tech' than has been the case so far, and in keeping with Dumont's views,

West Africa could benefit from the development of small-scale cottage industry geared solely to the processing of locally produced resources – sugar cane into sugar for example.

The impact of oil

No discussion of West Africa could be complete without some mention of Nigeria's oil. Oil has been known to exist in the delta region since colonial times, but it was only in 1956 that the Shell-BP Petroleum Development Company discovered oil in commercial quantities. This was the first of many finds in southern Nigeria. There are other reserves in West Africa, but no others as large as Nigeria's oil fields are yet known. Côte d'Ivoire's fields proved less extensive than had originally been hoped, Benin produces a limited quantity and exploration is now continuing along the Senegambian coast, and in Guinea-Bissau. Nigeria's oil is of excellent quality, low in sulphur and tar, but the reserves are finite and, at current rates of usage, are expected to last little more than another twenty to thirty years. Crude oil production rose dramatically from 3.4 million tonnes in 1962, to 114.2 million tonnes in 1979. By 1985, it had again fallen to 72.8 million tonnes. In 1978–79, the heyday of Nigerian oil, petroleum exports contributed over 90 per cent of export revenue. The impact of oil was astronomic and so was the generation of money in the country. Since 1981 and the squandering of resources by the civilian government, oil has declined in importance in the economy.

Oil brought vast riches to Nigeria, but its contribution to economic development was far less than it might have been. First, it changed the economy from one dependent on several resources, to a vulnerable 'monoculture'. This need not have happened, but with the abundance of money concentrated in the hands of a few powerful people, there was little incentive to maintain and develop other sectors and one that suffered heavily was agriculture. Oil has created employment opportunities both in exploration and in the development of the associated petrochemical industry, but only for some 16,000 people. Being a capital intensive industry, its capacity to use labour is limited. It has also had limited effects on other aspects of Nigerian industry. Backward linkages have been limited as there are so few industries in Nigeria making capital goods needed by the oil industry, and the capacity for forward linkages is similarly limited. Petroleum has generated funds rather than services, and while this has increased government revenue and expenditure, it has led to overspending and borrowing, the growth of inflation, a severely overvalued currency and all the problems associated with it (Rimmer 1985). A major benefit of oil to Nigeria is that its refineries have decreased demand for imports of petroleum and associated products since the mid-1960s. Although the oil industry is now slowing down, Nigeria does have considerable reserves of natural gas and methods of developing these are under way. Having squandered the benefits of oil, Nigeria is fortunate in having yet another valuable resource – gas – which could contribute significantly to the nation's development.

AGRICULTURE

The problems of transferring growth potential into reality

Agriculture is as important to West Africans now as it was at independence. Macro-economic indicators suggest that agriculture has declined relative to the service sector (Table 4.1), but in absolute terms it is still of crucial importance involving over 70 per cent of the population of most states. Many factors have caused change in agriculture. It is not possible to consider them all, so the following discussion focuses on three separate, but important causes: drought, population growth and major changes in government policy towards agriculture during the independence era. These allow some discussion of change in the agricultural sector at different scales.

Drought has been particularly severe in the northern states with vast losses of both crops and animals. However, recovery from devastation has been remarkable. Many semi-nomadic herders have moved to towns to find alternative forms of subsistence on either a temporary or permanent basis, but many have continued their tradition of herding and have since intensively restocked their herds. The result is that animal numbers have now passed 1970 levels in the Sahelian region (Toulmin 1990). Periods of drought have not been restricted to a few years, rather there has been a trend of lower rainfall since the late 1960s (Table 4.2), with severe shortages of water in some years. Farmers in the Senegambia and in Mali, for example, have adapted by obtaining more off-farm employment as a hedge against starvation if crops fail; they have made changes in the nature of land preparation in order to conserve soil moisture, they now sow quicker maturing varieties of millet, sorghum, groundnuts and rice, and in addition, they have been making greater use of bush resources – where they are available – particularly leaves, fruit and roots, in order to supplement dietary needs. Environmental change has thus brought about marked change in farming at field level.

In the north in particular population increase has induced significant changes in agricultural systems. Traditionally, pressure on land in the savannahs has not been great and local farming, characterized by low capital inputs, has been sustainable because four or five years' cultivation was followed by a lengthy fallow period during which the land could recover. With populations increasing, more land is being cultivated and essential fallow periods are being reduced to a minimum, or eliminated altogether. In the savannahs of the Senegambia, Northern Ghana and Nigeria, land which was left fallow for around six years a decade ago is now farmed every year. This is one of the most important changes in traditional farming in recent decades and could severely damage the environment and directly affect agricultural production in the future.

In parts of the forested south population pressure is extreme, and in Eastern Nigeria, for example, population pressure is now so great that over-cultivation is a major problem. Soil quality on farmland is not being maintained mainly because resources are not available, and gullying and erosion occur rapidly and are increasingly widespread. Population increase during the past three decades is leading to over-exploitation of the land and to the acceleration of destructive, positive feedback forces in the environment. It is vital that these are averted by changes in land management techniques.

Another aspect of the population problem is that as numbers grow and land is cleared for cultivation, trees have to be cut down. They are also felled for fuelwood for rapidly growing urban areas. While clearing bush land is contributing to environmental degradation, it also indirectly affects agriculture. A valuable input into the village comes from the forest or 'bush' where mainly women gather firewood, nuts, fruits, leaves, roots, fungi and other products which have a whole variety of uses as foods, herbs, medicines and dyes among other things (Siddle and Swindell 1990). But this natural resource is declining thus making food shortage or famine a very real possibility when harvests are poor.

The impact of government policy on the agricultural sector focuses the discussion at yet a different scale. In the 1960s agriculture was in a healthy state in many nations. Ghana, for example, earned considerable foreign exchange from cocoa and was a world leader in the crop's production. But the attitude of the Ghana government was that revenue from agriculture should be used to finance growth in other sectors of the economy, particularly industry. Low farm prices, lack of government support and the impact of overvalued currencies have demoralized farmers. When disease struck, or yields fell because cocoa trees were too old to produce, farmers were reluctant to replant, with the result that cocoa production declined in several West African states. Low prices and low investment in agriculture was not a feature only of the southern states, and throughout West Africa farming has been underachieving largely due to the low priority of farming in post independence development strategies (World Bank 1989: 19).

While the end of subsidies and price controls is intended to stimulate agricultural production for local consumption, currency devaluation is also intended to stimulate production of export crops. Promotion of export crops has had differing effects throughout the region. In Ghana for example, cocoa production has been increased again, but unfortunately earnings from cocoa have been lower than anticipated due to oversupply on the world market. In Nigeria marketing boards were eliminated in 1987. The immediate impact was a significant increase in farm prices for cocoa paid by independent cocoa producers. On the negative side, however, quality control was considerably more lax and Nigerian cocoa in particular rapidly received a bad reputation for its quality on the world market. The government had to step in reintroducing licences for cocoa merchants and inspectors to check on the quality of exports. Another example is from Mali which promoted cotton production to such an extent, that in 1985, as a complement to an excellent cotton crop, Mali also experienced food shortage (Timberlake 1985). Rubber production in Nigeria has also responded to the deregulation of the economy, and thousands of rubber trees in the Delta area of Bendel State are now being actively tapped again as world market demand for natural latex has risen. (The AIDS crisis is boosting condom sales world-wide, and at present these cannot be made with synthetic rubber.)

The search for appropriate forms of agricultural development

Policies towards rural development have varied considerably in space and in time and the aim here is to select specific examples of development schemes, and also to identify changing trends in agricultural development policy. Agricultural development did not have the highest of profiles in the strategies of West Africa's new leaders. However, as the catch phrase 'rural development' took hold of the developing world (Heyer et al. 1981), West Africa too began establishing projects to promote rural development. There were several in northern Nigeria: the Kano River Project, the Sokoto Rima and the South Chad irrigation project were some of the earliest, and in Senegal, irrigation projects along the Senegal River were greatly extended, though a rice-producing scheme had been started on the Delta during colonial times. There were similar projects in southern Senegal, at Jahaly Pachar in The Gambia and numerous others. They all shared certain characteristics: they were large scale and capital intensive, intended vastly to increase production of food crops, particularly rice, maize and wheat both for local consumption and for urban markets. They were inspired by national governments in conjunction with foreign aid agencies and feasibility studies were carried out. In spite of dubious results, USAID, FAO and the Commonwealth Development Corporation gave their blessing to the new Nigerian schemes in the early 1970s and finance was provided by other international agencies (Andrae and Beckman 1985: 77–86).

Almost invariably these schemes failed to meet their goals. Crop output fell below estimates while expenditure exceeded them. The schemes usually involved the use of irrigation and high-yielding varieties of crops, and farmers on the schemes frequently had little scope for decision-making. They were obliged to follow instructions from government agencies on when land should be ploughed, fertilizer applied and various other farm activities carried out (Baker 1982). Land was frequently prepared by the agency using foreign machinery; the seed was supplied by the agency and was not available on the open market (legally); fertilizer was in many instances imported, and a great deal of the agronomic expertise came from abroad. Indigenous skills and knowledge of the environment went to waste under these schemes. Low agricultural prices and highly restricted marketing methods gave farmers little incentive to put their skills into operation and to make the schemes work. Few large-scale schemes have succeeded in the way that was intended. The causes of problems have been well documented (Baker 1982; Adams 1985; Andrae and Beckman 1985, among many others), and there is little need to reiterate them here other than to say that without doubt, the mismanagement of human and material resources has left a legacy of social, economic and environmental problems. Alongside these shortcomings, large-scale schemes benefited many. The schemes created vast opportunities for patrimonialism to flourish, for job opportunities to be extended, particularly in project administration, and they also yielded valuable revenue to foreign firms from whom machinery had to be purchased and who therefore had to supply all spare parts. Numerous foreign advisers and consultants attached to the projects also benefited.

In the mid-1960s and 1970s such projects were politically prestigious and their existence, though not necessarily their success, seemed to epitomize national progress. It is interesting that small-scale schemes at the same time received very little government recognition. One such example was the work of the Chinese in the Senegambia, in Mali and in Mauritania. Expertise brought first by the Taiwanese and later by Chinese from the People's Republic was adapted to local conditions so that innovations were both ecologically sound, and well understood and accepted by the local people (Baker 1985).

There are strong arguments for and against large-scale, capital intensive, agricultural development and if the efficiency of large-scale schemes could be improved there is little reason why they should not succeed (Hart 1983). However, the trend has turned very much from large-scale to small-scale development schemes such as the introduction of shallow bore holes and hand pumps in fadamas of Northern Nigeria (Carter 1989). In Nigeria, for example, the government has withdrawn funding from the former large-scale schemes mentioned above, and is focusing on the Agricultural Development Projects (ADPs). Here the aim is to improve rural infrastructure, extension services, supplies of rural credit, and inputs – particularly improved seed and fertilizer to local farmers. ADPs are far from without problems, but their production potential appears far greater than on those schemes where farmers were little more than labour. Furthermore, there are greater hopes for ADPs with the benefits that Structural Adjustment is supposed to bring to the rural sector. There must be a note of caution about Nigeria's ADPs which are aiming to increase farm output through the use of improved seeds and irrigation, where possible, and fertilizer. In many of the ADPs cultivation used to involve a lengthy fallow period and in recent years this has been reduced significantly because of growing pressure on land and also because of growing availability of inorganic fertilizer. Over-cultivation is thus becoming a problem in some areas, and one that could ultimately result in any improvements in output being unsustainable. Considerable attention needs to be focused on the ecological suitability of modern agricultural development. The changing trend towards small-scale development schemes is, however, no guarantee of success.

As problems of environmental degradation increase, there is evidence that agricultural development in institutions such as the International Institute for Tropical Agriculture (IITA) and other centres are focusing on ecologically sustainable methods. Combining row and tree crops in certain forms probably offers one of the greatest possibilities for sustainable agricultural development so far, though such schemes are still relatively limited in their extent. There is also an increasing trend towards a holistic approach to agricultural development integrating agro, sylvo and pastoral systems. Even from the few examples cited above there is evidence of marked change in approach to agricultural development: with failures abundant, even the World Bank is now of the opinion that Africans could contribute usefully to new development initiatives. How these intentions are to be put into practice awaits to be seen.

CAUSES AND EFFECTS OF ENVIRONMENTAL DEGRADATION

Land is one of West Africa's greatest resources and over the past thirty years, for a variety of reasons land has been going out of production. The changing climate is partly to blame, and further dessication of Lake Chad is evidence of this. However, soil erosion and gullying in the south, or problems of salinity and water logging cannot be attributed to a changing climate, and are more the product of human use of unsound ecological practices. This section investigates some of the forces that appear to be bringing about land degradation, and it is evident that few are recent in their origin. It is just that their impact has never been felt as severely as it is now.

Drought has been a major contributor to environmental degradation in West Africa. It is not a new phenomenon and historical data support scientific evidence that West Africa has been in the throes of a progressive drying phase since about 5000 BP. Figure 4.2 (compiled by Nicholson 1981) shows the general trend in West Africa since the end of the fifteenth century (Farmer and Wigley 1985: 62). Over much of the past twenty five years there has been a

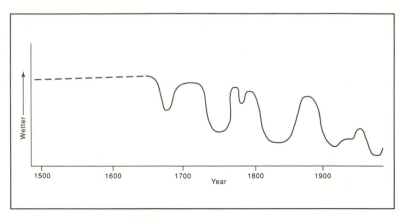

Figure 4.2   Summary of long-term rainfall changes in the Sahel and Sudan
regions of West Africa
Source: Farmer and Wigley (1985).
Note: Data prior to 1990 are largely qualitative, from historical documentary
sources (from Nicholson 1981).

general decline in the rainfall recorded within the region. Not only
have there been years of extreme drought such as 1968–73 and
1983–84, but even in relatively wet years the total rainfall has often
been less than expected due to a contraction of the rainy season. In
The Gambia for example, farmers are now geared to a rainy season
some five to six weeks shorter than it was in the mid-1960s.

Numerous hypotheses have been put forward to explain the
causes of the recent droughts and the current dry phase. One of the
more interesting discussions has focused on the role of sea surface
temperatures (SSTs) in climatic change. The seas of the South
Atlantic have been warmer than usual since 1968 with the North
Atlantic colder than usual. While this does not relate directly to
climatic anomalies in West Africa, Farmer and Wigley (1985) note
that it seems physically realistic to expect the persistence of such
large-scale differences to affect tropical climates. Work by Lough
(1981) has shown that years of lower rainfall in the Sahel are
correlated with warmer than usual SSTs in the Atlantic south of
West Africa. Explanations for the relationship are limited though it
has been suggested that a thermally driven zonal circulation which
is weaker, but similar to that in the Pacific, is believed to exist
between the cold waters of the Gulf of Guinea and the South

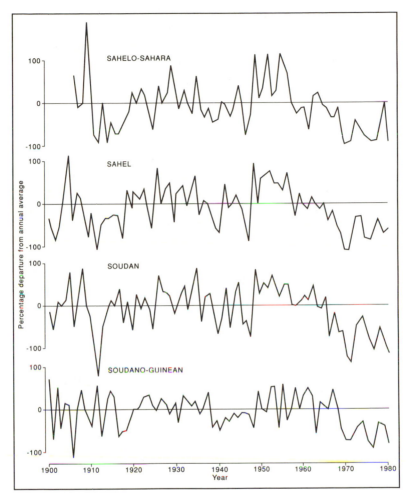

Figure 4.3   Rainfall departure from the annual average for four West
Africa zones 1990–80
Source: Farmer and Wigley (1985)
Note: Plotted values are approximately equivalent to a regionally averaged
'per cent or standard departure' (from Nicholson 1983).

American zone of strong convection. This Atlantic system also
shows quasi-periodic fluctuations, which it is thought might cor-
relate with rainfall in West and other parts of tropical Africa though
there is as yet insufficient evidence to support this.

Whatever the causes of drought, its effects have been greatest in the north of the region, but more indirect further south. For example, rivers in Côte d'Ivoire which flowed through drought areas were so low that the waters were insufficient to turn the turbines for HEP plants which affected domestic consumption, industry and energized tube well agriculture. In the north, shortage of water in drought years has resulted in severe losses of crops and animals and in food shortages. Successive years of drought and reduced rainy seasons have left the land exposed and vulnerable to erosion by wind and even to water – when it has arrived. Top soil has been blown and washed away and problems of desertification have become acute in these marginal lands. Loss of top soil has meant that even if rainfall has been adequate in some years, the capacity of the soil to produce has fallen, and low germination and growth rates have left the land open to further erosion. Drought has thus added to environmental degradation by promoting positive feedback mechanisms in the environment. Human action has also brought about degradation and the question remains how far this may ultimately be causing a change in climatic conditions, or how far the changing climate is due to 'natural circumstances'. The remainder of this section focuses on human impact on the environment.

Environmental degradation is not limited to the north, but historical factors have contributed more to the degradation of the north than the south. For example, by 1400, trans-Saharan trade was flourishing and twelve thousand camels a year were crossing the Sahara via the Ahaggar route (Gritzner 1988: 57). Charcoal was a very important item of trade and camel caravans were reluctant to cross the desert without it. The result of this was massive forest clearance and species such as *Acacia tortilis* and others suitable for the production of charcoal, were severely reduced. The gum arabic trade which operated between the eleventh and twentieth centuries, also contributed to the region's deforestation and degradation. Gum arabic which was used in the textile industry in Europe is stored by the *Acacia senegal*, and its production increases with stress. It thus became the practice to stress trees in the forests of Senegal to increase the yield of gum. But this proved counter productive in the longer term as *A. senegal* has a relatively short life and increasing stress on trees reduced their ability to fight disease. During the twentieth century the forests of *A. Senegal* have disappeared leaving considerable areas open to erosion by wind and water. In addition, *A. senegal* has an extensive root system. The deep tap root exploited moisture at depth, and the extensive surface rooting system was able to make use of surface moisture (Gritzner 1988; Grainger 1982). The roots thus had a stabilizing effect on the soil and their disappearance facilitated soil erosion.

Population growth has increased pressure on the region's fuel-wood resources (Allen and Barnes 1985). Problems are undoubtedly worse in the savannahs and Sahel region as wood is still more abundant further south. In the area around Bamako in Mali, for example, acres of tree stumps ar all that remain of forests cut for sale to a rapidly growing urban population either as wood, or as charcoal. Conservation efforts are now increasing throughout the region, but with the exception of some cities in Nigeria where attempts are being made to provide people with butane gas, the choice of alternatives for fuelwood remains very limited. Increasing work is being done on the possible expansion of both exotic and indigenous tree species in order to meet a variety of needs.

One traditional practice which is causing increasing environmental degradation is burning. Traditionally burning has been an essential practice in the management of savannahs (Cole 1986; Stott 1991; Webster and Wilson 1966). In West Africa there are several reasons why land is burnt: it is one way of thinning out the vegetation which has covered the plot since the previous season's harvest; if the burn is timed carefully the ash can act as a fertilizer for the next crop; the fire may be started intentionally by pastoralists wishing to promote the growth of grass with the rains; or it may simply be an accident. The traditional farming system involved cultivation of a piece of land for around five years after which it was left to fallow for as much as twenty years or even more. There are stilll parts of the region where land is abundant such as in Guinea and north-west Côte d'Ivoire and where long fallows form part of the farming system. But where population densities are higher as in the Senegambia, the Mossi Plateau in Burkina and in the zones around Nigeria's northern cities, fallow periods have been pro-gressively reduced and increasingly are being eliminated. As the land in cultivation is almost invariably burnt before the new crop is sown, so the land where the fallow has been eliminated is being burnt every year. This is more than the land can stand and the effects of fire are degrading the environment. Trees of economic value preserved on the plot can rarely stand more than three fires in successive years. If the fire is toward the end of the dry season it burns at a much higher temperature and sterilizes the soil destroying fauna, plant roots and what little organic matter there may be. Furthermore, the potential of the ash to act as fertilizer is more often than not lost because it is blown away particularly where the landscape is relatively open. Crop yields decline as there are insufficient inputs to prevent positive feedback mechanisms from coming into play. When land is left fallow or abandoned after it has been over-cultivated for many years and regularly burned, the return to the dominant type of vegetation tends to be a very slow process with the land being capable of supporting little other than grass species for at least a decade. Fire is thus one of the more important features of the rural scene which can hasten environmental degradation and its impact should not be underestimated.

Many of the causes of environmental degradation so far mentioned are not new, but some aspects of modern economic activity and development may also be environmentally hazardous. One problem with some large-scale irrigation schemes in northern Nigeria such as the Bakalori is that injudicious use of water has resulted in water logging and salinity problems which leads to land going out of production. Evidence is still limited but it has been suggested on capital intensive agricultural schemes such as the early ADPs in northern Nigeria, that greater use of fertilizer in these areas than elsewhere, is feeding into the water supplies and causing problems both for humans and for wildlife. Other aspects of modern economic development such as the use of tractors in areas where soil erosion is prevalent causes problems as does the use of heavy machinery in logging. In Côte d'Ivoire, for example, serious soil erosion has been promoted as logs are dragged by machinery and the landscape is torn apart leaving it open to erosion.

As land becomes an increasingly scarce resource imbalances in the environment will inevitably appear. There is growing awareness of the need to combat environmental problems but success is more likely if funding, technology and expertise from the West are combined to a very considerable extent with local skills, knowledge and expertise. There is talk of this happening, but as yet little hard evidence to show that it is.

CONCLUSIONS: THE LIMITS OF KNOWLEDGE

Writing this chapter has reinforced awareness that we still know so little about West Africa. Much use is made of macro-indicators, and they have been used here simply because no alternative exists. Their

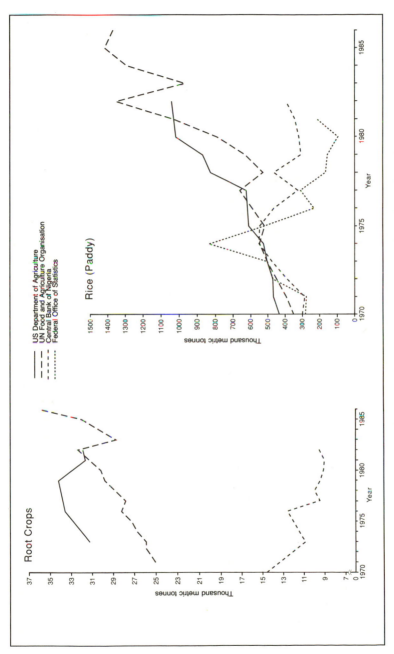

Figure 4.4 Production estimates for root crops and paddy in Nigeria
Source: Baker (1989: 8).

accuracy cannot be taken for granted, particularly those involving per capita data when we have no more than estimates of the number of people in the region. Thus reasoning from a statistical base has limited value in West Africa. Just one example of how data conflict is apparent from Figure 4.4. This shows the production of root crops in Nigeria and estimated paddy production with data from three different sources (Baker 1989). Clearly it would be hard to make any sensible conclusions about the nature of root crop production when the data vary so markedly. This is not to criticize Nigerian data collection methods; the example is used simply to demonstrate the spurious nature of statistics relating to the region and the difficulties of making sense of them.

On the more qualitative side, there is also much we do not know about West Africa, or perhaps it is known, but its importance has never been realized. The role of women has been highlighted only in the past fifteen years or so. And similarly, the role of the informal sector has been known, but its contribution greatly underestimated until about a decade ago. There is at last a growing awareness of the importance of the environment which hints at an appreciation of the need for an ecological approach to development. But neglect of the environment is no longer possible as increasing population pressure means than much more is being extracted from the environment than in times past.

The application of ecological principles to development schemes, particularly those in rural areas, could prove successful, but what is needed are African scientists with a knowledge of African environments, capable of incorporating indigenous skills and knowledge into scientifically sound development initiatives. There is an acute shortage of trained scientific personnel in West Africa and this is a gap that needs to be filled with Western assistance. Persistent poverty and low living standards will be obliterated only if the African input into solving West African problems increases significantly. The potential for positive change is there, but it seems unlikely to be realized in the near future.

REFERENCES

Adams, W.M. (1985) 'The downstream impact of dam construction: a case study from Nigeria', *Transactions, Institute of British Geographers* 10 (3): 292–302.
Allen C. (1989) 'The Congo', in Allen, C., Radu, M.S., Somerville, K. and Baxter, J. *Benin, The Congo, Burkina Faso*, London: Pinter.

Allen, J.C. and Barnes, D.F. (1985) 'The causes of deforestation in developing countries', *Annals, Association of American Geographers* 75 (1): 163–84.

Andrae, G. and Beckman, B. (1985) *The Wheat Trap*, London and Uppsala: Zed Books.

Araka, J., Ephson, B. and Horst, S. (1990) 'Farmers adjust to economic reforms', *African Farmer* 3: 5–15.

Awassi Atsimadja, F. (1992) 'The informal sector in Gabon', PhD thesis in draft, London: School of Oriental and African Studies.

Baker, K. (1982) 'Structural change and managerial inefficiency in the development of rice cultivation in the Senegal River Region', *African Affairs* 81 (325): 499–510.

Baker, K. (1985) 'The Chinese agricultural model in West Africa', *Pacific Viewpoint* 2: 401–414.

Baker, K. (1989) *Agricultural Change in Nigeria*, London: John Murray.

Bates, R.H. (1983) *Markets and States in Tropical Africa*, Berkeley: University of California Press.

Blaikie, P. (1985) *The Political Economy of Soil Erosion in Developing Countries*, London and New York: Longman.

Carter, R. (1989) 'The development of small scale irrigation in sub-Saharan Africa', *Public Administration and Development* 9: 543–55.

Chazan, N. (1983) *An Anatomy of Ghanaian Politics: Managing Political Recession, 1969–1982*, Boulder, Col.: Westview.

Cole, M.M. (1986) *The Savannahs: Biogeography and Geobotany*, London: Academic Press.

Collier, P. (1990) 'Africa: from crisis to sustainable growth', background paper no. 32, mimeo, for the Africa Development Group, School of Oriental and African Studies, London.

Cruise O'Brien, D.B., Dunn, J. and Rathbone R. (eds) (1989) *Contemporary West African States*, Cambridge: Cambridge University Press.

Dixon, R.A., Olanrewaju, D.O. and Shakur, T. (October 1989) *Health and Living Conditions of the Urban Poor in Nigeria: A Review*, mimeo, Centre for Development Planning Studies, University of Sheffield.

Dumont, R. (1982) *False Start in Africa* (translated from the French), London: Earthscan (first published in 1962 as *L'Afrique noire est mal partie*, Paris: Editions du Seuil).

Dumont, R. and Mottin, M.F. (1980) *L'Afrique etranglée*, Paris: Seuil.

Dunn, J. (ed.) (1978) *West African States: Failure and Promise*, Cambridge: Cambridge University Press.

Economist Intelligence Unit (1986–7) *Country Profile: Niger, Burkina*, London: EIU.

Farmer, F. and Wigley, T.M.L. (1985) *Climatic Trends for Tropical Africa: Research Report for the Overseas Development Administration*, Climatic Research Unit, School of Environmental Sciences, University of East Anglia.

Faure, Y.A. (1989) 'Côte d'Ivoire: analysing the crisis', in Cruise O'Brien, D.B., Dunn, J. and Rathbone, R. (eds) *Contemporary West African States*, Cambridge: Cambridge University Press.

Gakou, M.L. (1987) *The Crisis in African Agriculture*, London: Zed Books.

Gleave, M.B. (1988) 'Population pressure in West Africa', Department of Geography, University of Salford Discussion Paper in Geography no. 34, mimeo.

Grainger, A. (1982) *Desertification: How People Make Deserts, How People can Stop and Why they Don't*, London: Earthscan.

Gritzner, J.A. (1988) *The West African Sahel: Human Agency and Environmental Changes*, Chicago: University of Chicago Press.

Harrigan, J., Mosely, P. and Toye, J. (1991) *Aid and Power: The World Bank and Policy-Based Lending*, London: Routledge.

Hart, K. (1983) *The Political Economy of West African Agriculture*, Cambridge: Cambridge University Press.

Heyer, J., Roberts, P. and Williams, G. (eds) (1981) *Rural Development in Tropical Africa*, London: Macmillan.

Hodd, M. (1986) 'Africa, the IMF and The World Bank', *African Affairs*, September: 35–43.

Hulme, M. (1989) 'Is environmental degradation causing drought in the Sahel? An assessment from recent empirical research', *Geography* 74 (322): 39–46.

Jeffries, R. (1989) 'Ghana: the political economy of personal rule', in Cruise O'Brien, D.B., Dunn, J. and Rathbone, R. (eds) *Contemporary West African States*, Cambridge: Cambridge University Press.

Kirk-Greene, A. and Rimmer, D. (1981) *Nigeria since 1970: A Political and Economic Outline*, London: Hodder & Stoughton.

Lane, C. (1990) 'The Franc Zone: monetary union in crisis?', background paper no. 27, mimeo, for the Africa Development Group, School of Oriental and African Studies, London.

Lewis, L.A. and Berry, L. (1988) *African Environments and Resources*, Boston, Mass.: Unwin Hyman.

Lough, J.M. (1981) 'Atlantic sea surface temperatures and weather in Africa', PhD thesis (unpublished), Norwich: University of East Anglia.

Ngom, T. (1989) 'Appropriate standards for infrastructure in Dakar', in Stren, R.E. and White, R.R. (eds) *African Cities in Crisis*, Boulder, Col.: Westview.

Nicholson, S.E. (1981) 'The historical climatology of Africa', in Wigley, T.M.L., Ingram, M.J. and Farmer, G. (eds) *Climate and History*, Cambridge: Cambridge University Press.

Nicholson, S.E. (1983) 'Sub-Saharan rainfall in years 1976–80: evidence of continued drought', *Monthly Weather Review* III: 1646–54.

Nicholson, S.E. and Flohn, H. (1980) 'African environmental and climatic changes and the general atmospheric circulation in late Pleistocene and Holocene', *Climatic Change* 2: 313–48.

Nigerian Economic Society (1986) *The Nigerian Economy: A Political Economy Approach*, London and Lagos: Longman.

O'Connor, A., (1983) *The African City*, London: Hutchinson.

Onyemelukwe, J.O.C. (1984) *Industrialization in West Africa*, London: Croom Helm.

Peil, M. (1990) 'Whose city? Rich and poor in Lagos', paper presented to African Studies Association UK Conference, Birmingham, 11–13 September.

Richards, P. (1983) 'Farming systems and agrarian change in West Africa', *Progress in Human Geography* 7: 1–39.

Richards, P. (1990) 'Local strategies for coping with hunger: Central Sierra Leone and Northern Nigeria compared', *African Affairs* 89 (355): 265–76.

Rimmer, D. (1985) 'The overvalued currency and over-administered economy of Nigeria', *African Affairs* 84: 435–47.

Ruthenberg, H. (1980) *Farming Systems in the Tropics*, Oxford: Clarendon.

Siddle, D. and Swindell, K. (1990) *Rural Change in Tropical Africa: From Colonies to Nation States*, Oxford: Basil Blackwell.

Stott, P. (1991) 'Recent trends in the ecology and management of the world's savanna formations', *Progress in Physical Geography* 15 (1): 18–28.

Stren, R.E. and White, R.R. (eds) (1989) *African Cities in Crisis*, Boulder, Col.: Westview.

Timberlake, L. (1985) *Africa in Crisis: The Causes, the Cures of Environmental Bankruptcy*, London: Earthscan.

Toulmin, C. (1990) 'Pastoralists in peril: Sahelian pastoralists and their problems', paper presented to African Studies Association UK Conference, Birmingham, 11–13 September, mimeo.

Wallace, T. (1981) 'The Kano River Project, Nigeria', in Heyer, J., Roberts, P. and Williams, G. (eds) *Rural Development in Tropical Africa*, London: Macmillan.

Watts, M. (1983) *Silent Violence: Food, Famine and Peasantry in Northern Nigeria*, Berkeley: University of California Press.

Webster, C.C. and Wilson, P.N. (1966) *Agriculture in the Tropics*, London: Longman.

White, R.R. (1989) 'The influence of environmental and economic factors on the urban crisis', in Stren, R.E. and White, R.R. (eds) *African Cities in Crisis*, Boulder, Col: Westview.

World Health Organization (1990) End August 1990 Global Status of reported AIDS to WHO.

World Bank (1989) *Sub-Saharan Africa: From Crisis to Sustainable Growth*, Washington DC: World Bank.

World Bank (1990) *World Development Report*, Washington DC: World Bank.

5

THE CHANGING GEOGRAPHY OF EASTERN AFRICA

Anthony O'Connor

INTRODUCTION

Eastern Africa is not a clearly defined entity, either for the people who live there or for outsiders. Those people in Ethiopia who are aware of the existence of Tanzania do not feel that they have anything more in common with Tanzanians than with the people of Nigeria or Zaire. It is not even totally obvious which countries should be included. Parts of Mozambique, Malawi and Zambia are very similar both environmentally and culturally to much of Tanzania. Burundi and Rwanda would probably have been included if their colonial history had not tied them in many ways to Zaire. And a civil war is raging in Sudan partly because its southern peoples have far more in common with the people of northern Uganda than with their compatriots in the north of Sudan. Somewhat arbitrarily, attention will here be confined to Tanzania, Uganda, Kenya, Ethiopia, Somalia and Djibouti (see Figure 5.1).

'East Africa' is a more commonly used term, and is generally regarded as comprising Kenya, Tanzania and Uganda (e.g. Morgan 1973). This is largely the result of colonial history, for not only did Britain administer all three but also they shared various common services such as railways and posts. These were maintained for some years after independence in 1961–3; and until its eventual demise in 1977 the East African Community at least nominally constituted a common market. The most densely settled parts of each country lie nearer to their shared borders than to other borders, and the external orientation of Uganda in particular is firmly eastwards

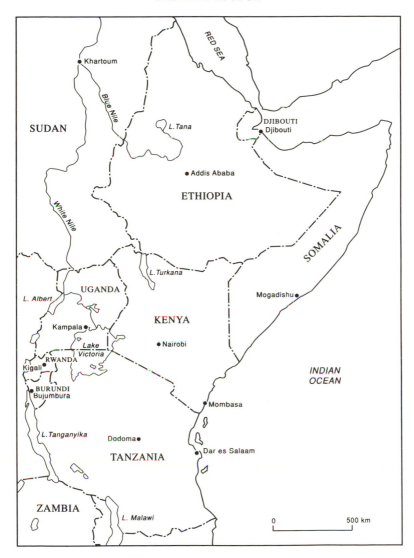

Figure 5.1 Eastern Africa

and southwards rather than westwards and northwards. Most importantly, many people feel some sense of 'East African' identity, despite recent changes such as Tanzania's membership of SADCC.

There is no such sense of common identity among the people of the region often labelled by outsiders as 'The Horn of Africa' (e.g. Markakis 1987), but political conflict has brought far closer interaction between Ethiopia and Somalia than between Ethiopia and Kenya. Indeed, many people living in the Ogaden region of eastern Ethiopia regard themselves as Somali, and war between the two countries in 1977–8 arose from Somalia's claim to this territory. Djibouti has somehow managed to avoid being swallowed up by either country, despite close cultural ties to Somalia and close economic ties to Ethiopia. For many people there is a seventh country in eastern Africa, as Eritrea was an Italian colony that has been administered by Ethiopia only since 1951, and as it is now largely controlled by those who have long fought for its independence. However, an independent Eritrea is not yet recognized in international law, and I shall refer to it here as part of Ethiopia. Similarly, in 1991 the northern part of Somalia was declared to be the independent state of Somaliland, though without international recognition.

For some purposes it is possible to discuss eastern Africa as a whole, although many of the common features are in fact common to the whole of tropical Africa. For other purposes it is more appropriate to consider the individual nation states, despite the arbitrary and externally imposed nature of the national boundaries. Ideally, perhaps, even smaller units should often provide the framework for discussion (Kesby 1977), especially as these may be far more meaningful to the people concerned, but this is not possible in a chapter of this length. Not many statements can be made about society, economy or environment in Ethiopia or Kenya that apply equally to all parts of either country. Within each country there is immense geographical diversity, but here the emphasis is on those features and trends about which some generalizations, at least at national level, can be made.

POPULATION GROWTH

The most fundamental change in the geography of eastern Africa over the past twenty-five years has undoubtedly been the growth in population, from around 50 million in 1965 to around 120 million in 1990. It is essentially because of this that more land is now cultivated, more food is produced and consumed, more economic activity of all types take place, and there is more pressure on the natural environment. By the 1980s even the building of new roads, new schools or new health centres was largely a function of population growth rather than real 'development' in the sense of improved welfare.

As in most parts of Africa, and in sharp contrast to much of Asia, the rate of increase shows no sign of slackening. Indeed, it has risen from under 3 per cent a year to around 3.5 per cent a year as death rates have fallen while birth rates have remained more or less constant. In so far as this change results from reduced infant and child mortality, and from people in general living longer, it is of course welcomed by families throughout the region, even though a 3.5 per cent growth rate severely constrains welfare improvements. Experience elsewhere, and the 'theory of demographic transition', suggests that falling death rates should be quickly followed by falling birth rates. At present, however, few parents in eastern Africa, women or men, wish to restrict the numbers of their children. The average fertility rate of six to seven children per woman is very close to the chosen number revealed in most surveys (Caldwell and Caldwell 1987). Children are seen as an economic asset, and as a form of security for old age, and to some extent as the main purpose of life. Where family planning programmes arouse any interest it is more in terms of spacing births, and even ensuring more, than in order to limit them.

There are of course some variations across the region, and Kenya is notable as the country with the highest rate of natural increase in the world, probably close to 4 per cent a year (Ominde 1984). In Kenya the death rate has fallen from twenty per thousand in 1965 to only eleven per thousand in 1988, while the birth rate has remained at around 50 per thousand. The death rate has fallen less in Tanzania, and much less in Uganda, while it remains at around twenty per thousand in Ethiopia and Somalia (UNICEF 1990). There are also spatial variations in birth and death rates within each country, and in certain years famine has been severe enough to have an impact in parts of Ethiopia. Variations in population growth rates within countries, however, are more often the result of migration. In national terms, massive refugee movements, across the Ethiopia–Sudan, Ethiopia–Somalia and Uganda–Sudan borders have more or less cancelled each other out, but they have brought great changes to particular districts. More generally, net rural–urban migration over the past twenty-five years has reduced the annual rate of increase to around 2 per cent in many rural areas, while producing an annual increase of 6 per cent to 9 per cent in many cities and towns. In this case there is thought to have been some slackening of the pace in the 1980s, but firm evidence on this is lacking. Much more certain is a variation from country to country, such as a much faster rate of urbanization in Tanzania than in Uganda during the 1970s (O'Connor 1988). And while rapid urbanization has been one of the most important changes occurring in eastern Africa, it is important to remember that over three-quarters of the population are still rural dwellers.

One consequence of rapid population growth is a demographic structure in which children make up almost half the total, putting an immense strain on the education system, and leading to a very high dependency ratio if they remain in school into their teenage years. The age structure also means that there is a huge built-in potential for further growth even in the unlikely event of a sharp fall in fertility. The rate of growth also means that promises such as 'health for all' or 'pure water for all' by the year 2000 cannot possibly be fulfilled. The absolute numbers of malnourished children, illiterate adults and people without access to clean water are in fact increasing year by year.

The attitude we take to this depends on whether we regard more people as in itself 'a good thing' – as most people in eastern Africa clearly do. In Kenya the government view is that population growth is too rapid, and that unless it can be checked the pressure on cultivable land will soon become intolerable: but the impact of its family planning programme has been very limited. Elsewhere, governments have shown very little interest. In Ethiopia, for example, the chosen method for dealing with regional population pressure is a programme of resettlement in more sparsely populated regions.

It is sometimes said that the appalling new disease, AIDS, is checking population growth in Uganda, Tanzania and Kenya. In fact up to 1990 its demographic impact at national level was negligible. Clearly, there is a grim possibility that HIV infection has been spreading so rapidly that within the 1990s death rates will rise steeply enough to reduce growth rates in all three countries. However, the fear of parents that some of their children may die of AIDS, either in infancy (if the mother is infected) or as young adults, may already be having the reverse effect by encouraging increased fertility.

So not only is rapid population growth of profound importance for most other aspects of life in eastern Africa, but also it is the one change occurring over the past twenty-five years that we can be sure is continuing into the early 1990s. In many other cases we know about past trends but can only speculate on whether these are persisting as present processes.

POLITICAL CHANGE

In the period 1960 to 1963 first Somalia, then Tanzania (as Tanganyika before the later incorporation of Zanzibar), then Uganda, and finally Kenya, attained independence. Since then there have been important political changes, but they have mainly been particular to individual countries rather than features of eastern Africa as a whole. Colonial rule persisted in one small corner of the region until 1977, when the French finally departed from Djibouti, leaving a micro-state whose continued independent existence has been rather remarkable (Laudouze 1989). Some would argue that a form of colonial rule remained within Ethiopia, involving not a European power but imperial rule by Addis Ababa over peripheral regions which were bitterly opposed to this (Cliffe and Davidson 1988).

Civil war has in fact been a feature of Ethiopia throughout the past twenty-five years, though its intensity has varied. Initially it arose through the people of the former Italian colony of Eritrea demanding independence rather than annexation by Ethiopia; but subsequently the people of Tigre, the Somali people of the Ogaden, and many Oromo groups elsewhere, also fought against control by the Amhara of the centre. By 1988 most parts of Eritrea and Tigre were effectively administered by the respective liberation fronts. The port of Massawa was taken in 1989, and Asmara, the Eritrean capital, finally fell in 1991.

The civil war was remarkably little affected by the great change in the Ethiopian political scene which occurred in 1974, when the imperial government of Haile Selassie crumbled and was replaced by a military regime. This declared itself Marxist-Leninist, turned to the Soviet Union for support, and greatly increased the power of the state. It gained the allegiance of some sections of the population, and it can probably be truly said that a 'revolution' has occurred in Ethiopia (Clapham 1987). However, regionally-based opposition to the regime led by Mengistu Haile Mariam intensified in the 1980s, and it was finally overthrown in 1991 by the Tigrean-led forces of the Ethiopian People's Revolutionary Democratic Front. The ideological position of the new government was not immediately clear, but it promised full independence for Eritrea if a referendum there favoured this.

Somalia has had no revolution, but it has experienced much political conflict as rival groups have competed for power (Laitin and Samatar 1987). In 1989 this conflict intensified, especially in the north, and it had spread throughout the country by 1991, leading to massive loss of life and huge displacements of people. A flow of refugees into Ethiopia resulted, counterbalancing earlier flows of Somali people from Ethiopia into Somalia. However, the movement of refugees from Ethiopia (including Eritrea) into Sudan has been even larger in scale, involving almost 1 million people (Bulcha 1988). Some have fled from war, some from the fear of being enlisted to fight against their own people, some from the famine to which the war contributed, and many from a combination of war and famine.

Yet another flow of refugees involved people from northern Uganda fleeing into southern Sudan and also into Zaire. This, however, represented only the tip of the iceberg of political disorder in Uganda that existed throughout the 1970s and early 1980s. In 1965 this was still a peaceful and well-administered country, with excellent prospects for economic and social development. However, it had inherited a strange constitution by which the old Buganda kingdom at the heart of the country had a large measure of autonomy. Divergence of interests led to the possibly unique situation in which the heartland of the country attempted to secede. The attempt was crushed, but the result was a government ruling with very little support from the Ganda, the largest and most central ethnic group. There was something of a power vacuum into which the army chief, Idi Amin, thrust himself in 1971. By the time he was ousted in 1979 tens of thousands had been murdered, the state had lost all credibility, and the economy was in semi-chaos. The removal of Amin led to a further power vacuum, and tyranny was replaced by anarchy with undisciplined soldiers rampaging through the country and with much further bloodshed. Fortunately, the new government established in 1986 by Yoweri Museveni is slowly establishing some sort of order to this shattered country (Hansen and Twaddle 1988; Rupesinghe 1989).

By comparison, both Tanzania and Kenya have enjoyed remarkable political stability and even an orderly replacement of one head of state by another. This owes much to the statesmanship of both Julius Nyerere and Jomo Kenyatta. The stability has permitted each country to work out a much more coherent ideological position than has ever been possible in Uganda. In Kenya this has been right of centre, with strong encouragement of private enterprise and more emphasis on economic growth than on equity (Miller 1984). In Tanzania the 1967 Arusha Declaration ushered in an attempt to build an African version of socialism, under the Swahili name of *ujamaa*. For twenty years the state attempted to organize people, with as little coercion as possible, into communal activity, and to take the place of private enterprise in the large-scale sectors of the economy. This all proved grossly overambitious (Hyden 1980), and the *ujamaa* dream is now fading: but it has brought one great change in the geography of the country by relocating well over half the rural population, at least 10 million people, from dispersed homes into nucleated villages. It is generally agreed that both the villagization programme and *ujamaa* policies more generally have done more harm than good to the economy, but that they have helped to spread social welfare more widely, and have probably contributed to the high degree of national integration that Tanzania has achieved (Hodd 1988; Yeager 1989). Kenya has become a little less poor, but there have been more severe political tensions there, which by 1990 were causing the government to be increasingly repressive.

If there is any common theme to be found in the evolving political scene in eastern Africa it is perhaps the immense importance of individual heads of state. In Ethiopia in the 1980s Mengistu Haile Mariam enjoyed as much personal power as Emperor Haile Selassie did in the 1960s. Tanzania followed a particular course because Julius Nyerere commanded infinitely more respect than any rival and he felt this course to be right. Kenya may have problems in store because Daniel arap Moi does not command quite as much respect as his predecessor, Jomo Kenyatta. Certainly it can be argued that Uganda has suffered both from the failure of Milton Obote to establish himself as a charismatic leader and from the success of Idi Amin in imposing his personality on all aspects of Uganda life in the 1970s. This means that a tremendous responsibility now rests on the shoulders of Yoweri Museveni in Uganda, as also in different ways on those of Julius Nyerere's successor, Ali Hassan Mwinyi, in Tanzania.

External relations

The East African Community dating from British rule had harmonized some aspects of economic management and service provision, but it came under increasing strain in the 1970s, largely because Kenya appeared to gain the lion's share of the benefits that it provided and because of the ideological differences between Kenya and Tanzania. However, the final break-up in 1977 also owed much to the chaos created in Uganda by Idi Amin. For some years the Kenya–Tanzania border was closed, and Tanzania developed stronger links with both Zambia and Mozambique.

Ugandan troops invaded north-western Tanzania in 1979, and this provided Nyerere with a pretext not only to drive them out but also to send his forces on to displace Amin from power. One year earlier the Ethiopians might have done likewise in Somalia after repelling a major invasion of the Ogaden if their supplier of weapons, the Soviet Union, had not dissuaded them.

The switch of Ethiopia's external relations from the United States to the Soviet Union after the 1974 revolution is the most significant change of this nature that has occurred in eastern Africa since 1965, though one might also note the simultaneous switch in the opposite direction by Somalia. No such change has occurred in Kenya, whose external relations have been very largely with the West ever since independence. Even Tanzania has never swung towards the Soviet Union, though it did establish some links with China in the 1970s, and has preferred to interact with countries such as Sweden rather than to remain too tied to Britain or to become too involved with the United States. The very limited involvement of any external powers in the political trauma of Uganda shows that not all Africa's troubles arise from outside interferences but the failure of Tanzania's efforts at 'self-reliance' demonstrate that despite political independence many other forms of dependence on the rich countries remain a persistent feature of eastern Africa.

CULTURAL AND SOCIAL CHANGE

A key feature of the cultural geography of eastern Africa is the way in which most people identify with an ethnic group smaller than the nation and generally associated with a specific, if not precisely bounded, territory. The word 'tribe' is commonly used in East Africa for these groups, and 'tribalism' is often seen as the main barrier to true national integration. In India or Brazil only a minority are regarded as 'tribal peoples', but most people in Kenya feel themselves to be Kikuyu or Luo or Kamba or whatever, at least as much as they feel themselves to be Kenyan. Language is a key factor maintaining this sense of ethnic identity: there are at least twenty quite distinct languages spoken in Uganda, none by more than a quarter of the population.

Most has been done to change this situation in Tanzania, where successful efforts since independence to build up a sense of national identity have included the promotion of Swahili as the common language, though even now not the home language of most people. One's 'tribe' is now much less important there than in either Kenya or Uganda. Indeed, Uganda's political traumas have probably increased most people's attachment to their own ethnic group. Similarly in Ethiopia, families may be more conscious of being Oromo or Gurage now than twenty-five years ago, although in the far north the attachment of some seems to be more to the ex-colonial entity of Eritrea than to a 'tribe' or language group.

Religion provides an alternative basis for differentiation, and in this respect there is much diversity among the countries of eastern Africa. Islam is totally dominant in Somalia, and also in Djibouti. In Tanzania, Islam and Christianity are roughly equally balanced, yet this has not created any great problem, perhaps because most people accept one or the other in combination with elements of indigenous religion. In both Kenya and Uganda, Christians far outnumber Muslims, but the putative adherence of Idi Amin to Islam brought this faith into prominence in Uganda for a while in the 1970s. The situation is more complex in Ethiopia, where the influence of the Orthodox Christian church has been greatly reduced by the revolution, but where opponents of the regime often clung tightly to their faith; and where the Eritreans have received some support from Islamic countries because Muslims form a higher proportion there than elsewhere – though not a majority as many outsiders believe.

Both Islam and Christianity have played important roles in education in eastern Africa, but the main change of the past twenty-five years has been the massive expansion of state-controlled school systems. In some ways this has been the greatest success story in terms of 'development', here as elsewhere in Africa. However, there are substantial national differences, while in general there was also a change from rapid improvement in school provision through the 1960s and 1970s to a much more static situation (except in absolute numbers) in the 1980s. It is also widely felt that reduced budgets for education in the late 1980s meant a serious decline in quality at every level from primary schools to universities.

The primary school enrolment rate rose between 1965 and 1987 from 50 per cent to 95 per cent in Kenya, and from 30 per cent to 65 per cent in Tanzania (UNICEF 1990). The relative increase was even greater in Ethiopia, from a mere 10 per cent to 35 per cent. The record in Somalia is very much poorer, the rise there being only from 15 per cent to 20 per cent. The broad pattern of change has been similar for secondary schooling, except that Tanzania has lagged behind at this level, concentrating on primary schooling and adult literacy instead. The official estimates indicate that the adult literacy rate there rose from 33 per cent in 1970 to 90 per cent in 1985. The equivalent rise in Kenya was from 32 per cent to 60 per cent, and that in Somalia from a mere 3 per cent to a still appalling 12 per cent. There has probably been a very substantial advance in Ethiopia, but widely differing figures are reported.

With respect to education there have been major changes in gender imbalances, especially in Tanzania. While the primary enrolment and adult literacy rates in the 1960s for females were only half those for males, today parity has very nearly been reached. The same applies in Kenya for primary enrolments, though the adult literacy rate in 1985 still showed a gap between 50 per cent for women and 70 per cent for men. In Ethiopia also great disparities remain, while in Somalia little if any equalization has occurred. In most respects there is little firm evidence of change over the past twenty-five years in the geography of gender. Women continue to play the dominant role in food crop production throughout East Africa, while men are largely responsible for livestock and for cash crops. The employment opportunities for women in the cities continue to be much poorer than those for men. There are many more women in city jobs now than in the 1960s, but there is also more competition for these. This is because the most obvious change, most notably in Tanzania, has been from male-dominated rural–urban migration to a flow involving men and women equally (O'Connor 1983). Even in Nairobi, where once there were twice as many men as women, the ratio has now fallen to 120:100. The common assertion that male migration is leading to an ever-increasing proportion of female-headed households in the rural areas is thus probably untrue. In so far as the burdens on rural women have increased in recent years, the main cause is that more schooling means that children can contribute less to tasks such as fuel-gathering and water-fetching – and even minding the younger infants.

Women have probably benefited at least as much as men from improved access to health care over the past twenty-five years. The gains may have been greatest of all for small children, partly due to extensive new immunization programmes. In Kenya in 1960 over 200 out of every thousand children died before the age of 5. By 1988 the figure had fallen to 110. In both Tanzania and Uganda child mortality was still about 170 per thousand in 1988, but this represented much improvement on the 1960 rate of around 240 per thousand. In Ethiopia, however, the reduction has only been from 290 to 260 (UNICEF 1990).

Another indicator of the availability of health care is the ratio of doctors to population. In this case the data produced by the World Health Organization indicate improvement between 1965 and 1985 only in Kenya, from one doctor per 13,900 to one per 10,000, and in Somalia, from one per 36,000 to one per 16,000. In Uganda the ratio is thought to have deteriorated from one per 11,000 to one per 22,000; while in Ethiopia the figure of one per 70,000 in 1965 had become one per 80,000 by 1985. Even in Tanzania this ratio has not improved, but an extensive network of rural health centres has been established, and the villagization programme has greatly assisted people's access to these. The main problem has been the very limited range of facilities that they can offer on a minute budget.

For many people in eastern Africa, of course, health care is normally a matter of whatever indigenous systems can offer (Good 1987). Very little information is available on how things have changed in this regard in recent years. Certainly there seems to have been little progress in integrating indigenous and 'Western' systems, as has been achieved in many parts of Asia. Possibly the greatest value of indigenous systems has been in responding to mental disturbance caused by the years of political turmoil in Uganda (Dodge and Raundalen 1987).

There is much speculation on its potential role also in relation to the current horrific spread of AIDS. As yet this has caused very few deaths compared with many other diseases, but in parts of Uganda, Tanzania and Kenya the proportion of people who are HIV-positive is as high as anywhere in the world, and rising rapidly (Miller and Rockwell 1988). The Kenyan government has done all it can to play down this issue, for it could all too easily lose both its foreign tourists and its role as host to international organizations: but survey findings indicating that 90 per cent of prostitutes in Nairobi are HIV-infected have been released. Uganda has less to lose by acknowledging the problem, and is probably exceptional in suffering a high incidence of AIDS in some rural areas as well as in the towns (Hooper 1987). The Uganda Ministry of Health was stating by 1990 that 10 per cent of all adults in the country, and over 20 per cent of all in the capital city of Kampala, might be HIV-positive. Where it has taken hold, AIDS is already imposing a massive health care burden, is causing intense suffering to the victims and their families, and is leaving increasing numbers of small children as orphans. During the 1990s it may take a heavy toll of the most active members of the labour force, and may even contribute to a fall in food production nationally as it already has done locally (Barnett and Blaikie 1992). It threatens to bring devastating changes to Uganda, Kenya and Tanzania over the next twenty-five years, though not, so far as we know, to Ethiopia, Somalia or Djibouti.

ECONOMIC CHANGE

The past twenty-five years have brought to eastern Africa much demographic, political and social change; but economic change has been more limited. Ethiopia continues to be one of the poorest countries in the world, with GNP per capita in 1988 recorded by the World Bank (1990) as only $120. It is generally thought that average per capita incomes there have stayed roughly constant since the mid-1960s, though for some regions and for many families there have been years of steep decline into destitution, and years of recovery. The picture is similar in Somalia, although the 1988 estimate for that country is rather higher, at $170. This is also the figure to which GNP per capita in Tanzania has recently fallen, as a result of devaluation, and there too per capita incomes are again at about their 1965 level. Urban incomes in particular rose in the late 1960s and 1970s, but fell back in the 1980s.

Only in Kenya are people in general clearly better off now than in 1965. Even there this does not apply to all families, but rising incomes in the 1960s and 1970s were not confined only to a small elite, as is sometimes suggested. There were real gains for many small farmers and for urban workers (Hazlewood 1979); and these were not all lost in the 1980s. In Uganda, by contrast, per capita income change between 1965 and 1988 was equivalent to an annual fall of 3 per cent – one of the worst records in Africa. The political convulsions of the 1970s and early 1980s brought devastation to the economy, and only small steps towards recovery have yet been made. In each country (apart from Djibouti) 70 per cent to 80 per cent of all families continue to gain their livelihood primarily from agriculture or livestock rearing, although the proportion is falling everywhere, and although most rural dwellers also engage in many other activities, from occasional house-building to daily water-fetching.

In Kenya the late 1960s and early 1970s brought a great expansion of coffee cultivation on small farms in the Kikuyu, Embu and Meru areas, after its long prohibition during the colonial period. Then tea cultivation, formerly confined to plantations, also extended to many high-altitude small farms. Meanwhile, many of the formerly European-owned large mixed farms were bought by the government for resettlement schemes. Where previously Kenyan agriculture had been highly dualistic, with almost no medium-sized farms, it now became much more diverse (Hunt 1984; Livingstone 1986), with large farms being split up into units of varying sizes, and with the more successful smaller farmers expanding their holdings by buying out their neighbours. More and more of Kenyan farmland was then being registered as privately owned freehold, rather than being held by the community for the use of individuals. Most of these changes had slowed down greatly by the 1980s, when the main changes were intensification in the well-watered areas and extension of cultivation into marginal areas, both in response to the rapid growth of population.

While the Kenyan government has been encouraging entrepreneurial small farmers, many of whom now hire labourers from near-landless families (Collier et al. 1986), the Tanzanian state has been discouraging this type of economic stratification (Collier et al. 1986). The original *ujamaa* concept anticipated a shift not only to living in nucleated villages but also to working on communal farms, but this proved very unpopular and had to be abandoned. It was hoped that shared use of mechanical equipment would be possible on these farms, but this rarely proved financially viable. For many years farmers were also discouraged from selling surplus crops by the very low prices offered by the state marketing boards. Rural Tanzanians have managed to feed themselves through the 1970s and 1980s (Boesen et al. 1986), but there is little sign of more widespread education and health care leading to increased agricultural productivity. Ugandans have also continued to feed themselves adequately, except when violent conflict destroyed the farms of part of the well-watered south and when drought hit the largely pastoral north-east. Some farmers in the south have also continued to pick and sell their coffee, although a large proportion of it has been smuggled out of the country from the mid-1970s onwards. However, the cultivation of cotton in the east and north has almost ceased, while tea production in the West has suffered greatly from the breakdown of transport and marketing systems.

The greatest disasters within the rural economies of eastern Africa have of course been the famines that have struck large parts of Ethiopia, and also parts of Somalia, and these will receive special attention later (see pp. 131–2). In general, however, the agricultural scene in Ethiopia has changed remarkably little over the past twenty-five years, despite the revolution and substantial changes in land-holding arrangements resulting from it. Over large areas the state has replaced private landlords, but the small farmers still operate in much the same way as before. In some well-watered southern areas, where coffee is a cash crop, there has been scope for increasing the amount of land cultivated as population has grown: but in some northern areas, with more erratic rainfall, intensification of cultivation on steep hillsides is bringing severe environmental degradation and falling crop yields.

In sharp contrast to countries such as Nigeria and Zambia, minerals are of little importance in the economies of eastern Africa. Many mineral deposits exist, but none is large or rich enough to provide a major source of income. Tanzania had hopes of using deposits of coal and iron ore in the southern highlands as a basis for industrialization, but nothing has come of this. As far as minerals are concerned the main change of the past twenty-five years has been the increased price that each country has had to pay for oil imports.

Manufacturing occupied a prominent place in the development plans of each country in the late 1960s and early 1970s. The early stages of import substitution had already been implemented very successfully, with local cotton used to produce textiles, local limestone used to produce cement, and (sadly) local tobacco used to produce cigarettes. This process continued for a while, but in general the 1970s and 1980s have brought great disappointment. Many of the new factories have operated at a loss, and most are now operating at far below capacity. In some cases the main factor is lack of effective demand for the products, as incomes have fallen; in others it is lack of foreign exchange for the purchase of essential imports or spare parts. Already in the 1960s industry was best developed in Kenya, and it is there that the greatest progress has continued to be made. In Uganda there has been substantial retrogression, while elsewhere the picture is one of stagnation.

There seems very little prospect of eastern Africa following the example of south-east Asia by developing export-orientated manu-facturing in the near future. It is often said that island Mauritius has shown what could be done, but economically as well as culturally Mauritius has far more in common with south Asia than with eastern Africa, despite its location. Countries such as Ethiopia and Tanzania simply do not have any comparative advantage that would enable them to compete in world markets for manufactures.

One respect in which eastern Africa does have such advantage is in the market for international tourism. Sunshine and warmth all the year, an attractive coast, and especially the world's greatest concentration of large wild animals, have brought tourists in increasing numbers to Kenya (Bachmann 1988). In years of low export prices for coffee and tea this has even become the chief earner of foreign exchange. There were similar opportunities in Tanzania, but there the government has had more qualms about investing in luxury facilities for tourists while its own people lack many of the most basic of needs. Uganda also was attracting foreign visitors to its wildlife parks in the 1960s: there the tourist industry has been yet another of the casualties of political disorder.

Whereas the hotels in and around Mombasa depend mainly on foreign tourists, those in Nairobi depend equally on foreign business people. The city has been chosen for the regional headquarters of many international companies and organizations, and even for the global headquarters of the United Nations Environment Programme and the United Nations Commission on Human Settlements. Addis Ababa, the Ethiopian capital, also performs international functions, for it houses the headquarters of both the Organization of African Unity and the UN Economic Commission for Africa. Administration, commerce and finance have provided the main bases of the urban economy in Dar es Salaam, Kampala and Mogadishu also, albeit only at national level, having been given a great boost with the transfer of functions from London and Rome at independence. These have also provided the main rationale for urbanization in most of the smaller centres of the region, with manufacturing playing only a subsidiary role almost everywhere.

The greatest changes in the nature of urban economies in the 1970s and 1980s have related not to the cities' functions but to the structure and scale of economic activity. In the early 1960s most people working in Nairobi, Dar es Salaam or Kampala were employed either by the state or by large firms, the main exception being retail trade which was dominated by small Asian-owned (people of Indian descent) enterprise. Employment in such large-scale organizations has increased far more slowly than the labour force, so an increasing proportion of workers are now self-employed or in very small-scale enterprises – often labelled 'the informal sector' (ILO 1972). The change has proceeded furthest in Kampala and other Ugandan towns, where formal sector employment shrank in the 1970s, and from where most Asian traders were expelled in 1972 by Idi Amin in a popularist and racialist ploy. The gap was filled by the *magendo* economy, which is 'informal' in every sense, and all technically illegal, but which may involve very large transactions. A substantial 'parallel economy' has also developed in Tanzania, partly due to the ineffectiveness of state enterprises and partly due to a huge deviation between official and unofficial currency exchange rates (Maliyamkono and Bagachwa 1990). Formal large-scale enterprise remains much more dominant in Nairobi, but even there the municipal buses are now supplemented by small-scale *matatu* minibuses; there is flourishing micro-scale manufacture of lamps and cooking stoves in the Shauri Moyo district, and there are traditional healers to supplement the official health services (Good 1987).

Housing is one sector in which the changed balance between 'formal' and 'informal' provision is particularly evident. At independence in 1963 almost everyone in Nairobi lived in housing built either by government or by large private firms, but soon afterwards squatting began in areas such as Mathare, and by the mid-1970s half the population was living in illegally built shacks (Hake 1977). Today fewer are true squatters: most are renting a room in a structure built as a small-scale business enterprise by one of the large emerging group of landlords. Squatting has also become very extensive in Dar es Salaam, but there too more families are renting a room in a new six- or eight-room structure than living in a dwelling that they have built for themselves. In this case there has been less change in Kampala because private small-scale house construction was already much more common there at independence.

Addis Ababa is quite different from the East African capitals, for it is an indigenous city which did not inherit a colonial economic structure. There, small-scale enterprise has always been dominant, whether in trade, in manufacturing, or in housing. In some respects the attempts to impose much greater state control of the economy have brought changes contrary to those in East African cities. Housing, for example, is much more strictly regulated now than twenty years ago.

In general, incomes in the small-scale or 'informal' sector are lower than in the large-scale or 'formal' sector, but the extent of overlap has been increasing as real wages in the large-scale sector have fallen. Some micro-businesses have prospered, but many others bring pitifully small returns. Often these constitute under-employment, not in the sense of short hours but in the sense of a dozen young men all trying to sell ball-point pens to the same group of passengers waiting at the bus terminal. Outright unemployment has also greatly increased recently, not so much among the very poorest who could not survive without some means of support, but rather among school-leavers who cannot find employment that matches their aspirations and who remain dependent on their families.

The austerity measures which the governments have increasingly been required to take as a condition of loans from the IMF, and which in most cases were really needed, have caused widespread suffering. But those who believe that there is always bias in favour of urban areas, and have therefore argued that poor rural dwellers have been hit hardest, are flying in the face of the evidence as far as eastern Africa is concerned.

FAMINE IN ETHIOPIA

Ethiopia is normally very largely ignored by the rest of the world, but this changed in 1973 as news of widespread famine began to emerge. Famines had occurred many times before in various parts of the country, but rarely had so many people been afflicted. The severity of the famine, and the failure of the government to respond to it, certainly contributed to the downfall of Emperor Haile Selassie, and his replacement by the Mengistu military regime.

In the later 1970s famine retreated, but by 1983 it had returned again (Hancock 1985; Goyder and Goyder 1988). This time more than 9 million people were directly affected by severe food shortage, many of them having to abandon their homes and to trek to the relief camps set up by the national Relief and Rehabilitation Commission (RRC) along the few motorable roads. This body appealed for international assistance, and estimated that 900,000 tonnes of grain would be needed in 1984 in order to prevent mass starvation. Countries with food surpluses and United Nations bodies were slow to respond, both because the Ethiopian government failed to give wholehearted support to the RRC and because it was thought that the country's ports could not handle such a volume of grain even if it were sent.

A BBC film shown in October 1984 finally alerted the world to the gravity of the situation, bringing a rapid response from voluntary organizations and then from governments in Europe and North America. By this time international bodies such as the World Food Programme had also been fully mobilized, and in 1985 over 1 million tonnes of grain were provided as emergency aid to Ethiopia (Gill 1986; Jansson et al. 1987). Port capacity was increased very effectively, but overland communications in many areas are so poor that large quantities of food had to be moved to where it was needed by air.

It is thought that well over half a million people, many of them small children, died during 1983–5 in Ethiopia at least partly due to lack of food. In many cases, of course, disease was also involved, exacerbated by the overcrowded, insanitary conditions in the relief camps and feeding centres. A further half a million people trekked into Sudan as refugees, in many cases fleeing from famine as well as from the civil war. Appalling as this record is, it would have been vastly worse if foreign assistance had never arrived. It did save millions of lives.

There has been much debate about the cause of the famine, some stressing political factors while others put all the blame on drought, some thinking essentially of immediate causes while others are concerned with underlying structures. Clearly, many factors are involved. A rapidly increasing population scratching a living from the land in areas that are experiencing environmental degradation must be seen as highly vulnerable. They could be plunged into famine either by severe drought or by civil war. Over large parts of Ethiopia in the mid-1980s people were faced with a deadly combination of the two. By 1986 famine conditions had largely abated, but vulnerability persisted, and millions of people remained partially dependent on a continuing inflow of relief food. By 1990 drought had once again returned both to areas controlled by the government, such as Harerghe and Wollo, and to most parts of Tigre and Eritrea which by then were under the control of the liberation movements. Harvests had once again failed, and once again millions of people were facing the threat of starvation.

ENVIRONMENTAL ISSUES

It would generally be assumed that the physical environment is the aspect of an area that would change least over a twenty-five-year period. Many features of the eastern African environment have

indeed changed little since 1965. Not only is the geology unchanged, but also so is our knowledge of it: no major new mineral
discoveries have been made, for instance. Landforms have not
altered, and there have been no water management schemes comparable to those which created the Aswan, Kariba and Volta lakes
elsewhere in Africa in the 1960s. Hydro-electricity schemes on the
Tana River in Kenya and the Rufiji River in Tanzania have been
modest in scale.

Much the most significant short-term environmental changes
have concerned rainfall. As across much of Africa, rainfall has
generally been lower in the 1970s and 1980s than in the 1950s and
1960s. Severe drought in large parts of Ethiopia in the early 1970s
contributed directly to widespread famine there, and this was
repeated in 1983–5. Many meteorological stations recorded well
under half the mean annual rainfall in 1984, and this had particularly
devastating effects because it followed three or four years of
relatively poor rains. The 1984 rains failed to a comparable degree in
large parts of Kenya, but there previous years had been better, while
the mechanisms for coping with food scarcity were far superior in
Kenya (Downing et al. 1989).

It is quite impossible to say whether any long-term climatic
change is taking place in eastern Africa, but there is no doubt that a
process of environmental degradation is occurring in many areas.
Soil erosion is extremely severe in the northern highlands of
Ethiopia, and as soils become thinner on the steep slopes more and
more of the rainfall washes down them, taking more soil with it,
rather than remaining available for plants. Parts of Kenya and
Tanzania also are suffering from increasingly impoverished soils,
and in some cases from related increasing aridity. Observers have
termed this 'desertification'. Accelerated soil erosion is partly due to
change that has been taking place in the vegetation cover, as ever
larger areas have to be cultivated each year. It is in this way that
the human impact on the environment has been most widely
experienced. In spite of conservation efforts (Anderson and Grove
1987), forests have been shrinking as, for example, on the slopes
of Mount Elgon in eastern Uganda. Over large tracts of Ethiopia,
trees have largely disappeared, not only because of clearance for
cultivation but also because of reliance on wood for fuel. Where
such fuel has become really scarce, people have had to turn
increasingly to cowdung as an alternative fuel, thereby robbing the
soil of a major input that helped to maintain fertility.

It would be quite wrong to imagine that the vegetation is changing dramatically throughout eastern Africa. For instance, vast areas remain covered by *miombo* savannah woodland, and will remain so for decades to come, for these areas are very thinly populated. Vegetation change is important mainly because it is greatest in those areas where most people live, and because it directly affects the lives of the majority of rural dwellers in the region. The need to walk further in search of fuel is one of the clearest illustrations of this. It also illustrates how many aspects of change in the region are interrelated (Hjort and Salih 1989), for it was expected that wood fuel would increasingly be replaced by kerosene until the sharp rises in oil prices ruled that out. Instead of switching to kerosene and electricity, as had been intended, those living in the new nucleated villages of Tanzania often continue to rely on wood fuel, rapidly depleting local supplies. Increased concentration of livestock in these new villages is also placing new strains on the local environment. Meanwhile, many people are suffering from greater depredations of their crops by wild animals now that they do not live so close to their fields.

Large wild animals are, of course, one of the most highly distinctive features of the eastern African environment (Yeager and Miller 1986), and one of the bases of the tourist industry mentioned above. All too little is known about how their numbers and distribution have changed over the past twenty-five years. In areas such as the Tsavo National Park in Kenya, and the Ngorongoro Conservation Area and adjacent Serengeti Plains in Tanzania, control measures have probably maintained numbers of most of the species. In some other areas, extensive poaching has taken place, particularly in Uganda where wildlife conservation has been one of the many casualties of political disorder, and where more and more people had to turn to illegal activities if they were to maintain their incomes.

CONCLUSIONS

It should be evident from what has been said that eastern Africa is an area of great diversity. In terms of national units it includes one of Africa's largest, Ethiopia, and one of the smallest, Djibouti. Tanzania has enjoyed remarkable political stability over the past twenty-five years, while Uganda has experienced turmoil. Kenya has achieved growth in several sectors of its economy, and much improvement in fields such as education, whereas this cannot be said of Somalia.

Even within each country there is huge diversity, and this chapter may not have done full justice to this. Parts of Kenya are highly productive, densely populated and rapidly changing, while other parts are arid, very sparsely settled and largely unaffected by change elsewhere in the country. For many people, to think of Ethiopia is to think of either famine or war; but the majority of Ethiopians have no direct experience of either of these. Even within tiny Djibouti there is a world of difference between the crowded port-city and the areas around it occupied by nomadic pastoralists.

There is also much diversity, both among the countries and within each, in the extent to which facts are known and trends are understood. Kenya produces each year a very comprehensive statistical abstract; in Uganda the equivalent series ceased in 1971. No one knows how much coffee has been exported since then, much of it having been smuggled out through Kenya. This, of course, means that the apparently precise Kenya export figures include an unknown amount originating in Uganda. And while many up-to-date figures are published for Kenya, it is notable that not even preliminary figures from the 1989 census had appeared by the end of 1990. The number of AIDS deaths may well be better known in Kenya than elsewhere, but there are plenty of reasons for not releasing such information.

The past twenty-five years have brought increased recognition of the importance of local people's knowledge of their own immediate environment; but this knowledge is often not sufficient to cope with changes such as the doubling of population within twenty years. This indigenous knowledge must somehow be linked with expertise from outside if effective environmental management is to be achieved in eastern Africa. In parts of Kenya, and in even larger parts of Ethiopia, the need for such improved management is becoming desperate.

At the same time, we should acknowledge that we do not even know whether population increase is beginning to slow down, or is likely to do so in the 1990s. If the rate of urbanization is slowing, rural population densities might be rising faster than ever. Our ignorance on such matters is profound. The arrival of AIDS further compounds this ignorance of what is happening demographically. The one thing that we do know is that throughout eastern Africa most parents want at least as many children as they are having, and that they have no regrets about the population growth that has been the most fundamental and widespread change over the past twenty-five years.

The greatest cause for regret has been the political conflict which has brought such misery in both Uganda and Ethiopia. Drought cannot be prevented, though ways of coping with it must be improved. Population growth is desired, though it poses great challenges and may in fact be curbed involuntarily through rising death rates. Political conflict, especially on a scale that contributes to the death rate, is neither desired nor inevitable. We must hope that Kenya and Tanzania can continue to avoid it, and that Uganda, Somalia, and Ethiopia may somehow find peace in the 1990s.

REFERENCES

Anderson, D. and Grove, R.H. (eds) (1987) *Conservation in Africa*, Cambridge: Cambridge University Press.

Bachmann, P. (1988) *Tourism in Kenya: a Basic Need for Whom?*, Berne: Peter Lang.

Barnett, T. and Blaikie, P. (1992) *AIDS in Africa*, London: Belhaven.

Boesen, J., Haunevik, K.J., Koponen, J. and Odgaard, R. (eds) (1986) *Tanzania: Crisis and Struggle for Survival*, Uppsala: Scandinavian Institute of African Studies.

Bulcha, M. (1988) *Flight and Integration: Causes of Mass Exodus from Ethiopia and Problems of Integration in the Sudan*, Uppsala: Scandinavian Institute of African Studies.

Caldwell, J. and Caldwell, P. (1987) 'The cultural context of high fertility in sub-Saharan Africa', *Population and Development Review* 13 (3): 409–37.

Clapham, C. (1987) 'Revolutionary socialist development in Ethiopia', *African Affairs* 86 (343): 151–65.

Cliffe, L. and Davidson, B. (1988) *The Long Struggle of Eritrea*, Nottingham: Spokesman.

Collier, P. and Lal, D. (1986) *Labour and Poverty in Kenya, 1900–1980*, Oxford: Oxford University Press.

Collier, P., Radwan, S. and Wangwe, S. (1986) *Labour and Poverty in Rural Tanzania*, Oxford: Oxford University Press.

Dodge, C.P. and Raundalen, M. (eds) (1987) *War, Violence and Children in Uganda*, Oslo: Norwegian University Press.

Downing, T.E., Gitu, K.W. and Kamau, C.M. (eds) (1989) *Coping with Drought in Kenya*, Boulder, Col.: Lynne Rienner.

Gill, P. (1986) *A Year in the Death of Africa*, London: Paladin.

Good, C.M. (1987) *Ethnomedical Systems in Africa*, Hove: Guilford.

Goyder, H. and Goyder, C. (1988) 'Case studies of famine: Ethiopia', in Curtis, D., Hubbard, M. and Shepherd, A. (eds) *Preventing Famine*, London: Routledge.

Hake, A. (1977) *African Metropolis: Nairobi's Self-help City*, London: Chatto & Windus.

Hancock, G. (1985) *Ethiopia: The Challenge of Hunger*, London: Gollancz.

Hansen, H.B. and Twaddle, M. (eds) (1988) *Uganda Now: Between Decay and Development*, London: James Currey.

Hazlewood, A.H. (1979) *The Economy of Kenya: The Kenyatta Era*, Oxford: Oxford University Press.

Hjort, A. and Salih, M. (eds) (1989) *Ecology and Politics: Environmental Stress and Security in Africa*, Uppsala: Scandinavian Institute of African Studies.

Hodd, M. (ed.) (1988) *Tanzania after Nyerere*, London: Pinter.

Hooper, E. (1987) 'AIDS in Uganda', *African Affairs* 86 (345): 469–78.

Hunt, D. (1984) *The Impending Crisis in Kenya: The Case for Land Reform*, Aldershot: Gower.

Hyden, G. (1980) *Beyond Ujamaa in Tanzania*, London: Heinemann.

International Labour Organization (ILO) (1972) *Employment, Incomes and Equality: A Strategy for Increasing Productive Employment in Kenya*, Geneva: ILO.

Jansson, K., Harris, M. and Penrose, A. (1987) *The Ethiopian Famine*, London: Zed Books.

Kesby, J.D. (1977) *The Cultural Regions of East Africa*, London: Academic Press.

Laitin, D.D. and Samatar, S.S. (1987) *Somalia: Nation in Search of a State*, Boulder, Col,: Westview.

Laudouze, A. (1989, 2nd edn) *Djibouti*, Paris: Karthala.

Livingstone, I. (1986) *Rural Development, Employment and Incomes in Kenya*, Aldershot: Gower.

Maliyamkono, T.L. and Bagachwa, M.S.D. (1990) *The Second Economy in Tanzania*, London: James Currey.

Markakis, J. (1987) *National and Class Conflict in the Horn of Africa*, Cambridge: Cambridge University Press.

Miller, N. (1984) *Kenya: The Quest for Prosperity*, Boulder, Col. Westview.

Miller, N. and Rockwell, R.C. (eds) (1988) *AIDS in Africa: The Social and Policy Impact*, Lewiston: Edwin Mellen.

Morgan, W.T.W. (1973) *East Africa*, London: Longman.

O'Connor, A.M. (1983) *The African City*, London: Hutchinson.

O'Connor, A.M. (1988) 'The rate of urbanisation in Tanzania in the 1970s', in Hodd, M. (ed.) *Tanzania after Nyerere*, London: Pinter.

Ominde, S.H. (ed.) (1984) *Population and Development in Kenya*, Nairobi: Heinemann.

Rupesinghe, K. (ed.) (1989) *Conflict Resolution in Uganda*, London: James Currey.

UNICEF (1990) *The State of the World's Children 1990*, Oxford: Oxford University Press.

World Bank (1990) *World Development Report 1990*, Oxford: Oxford University Press.

Yeager, R. (1989, 2nd edn) *Tanzania: An African Experiment*, Boulder, Col.: Westview.

Yeager, R. and Miller, N. (1986) *Wild Life, Wild Death: Land Use and Survival in Eastern Africa*, Albany, NY: State University of New York Press.

6

THE CHANGING GEOGRAPHY OF NORTH AFRICA

Development, migration and the demographic time bomb

George Joffé

INTRODUCTION

Between 1951 and 1962 the five states of North Africa – Libya, Tunisia, Algeria, Morocco and Mauritania – gained their independence and individually began what has proved to be a very difficult road towards economic development and modernization (see Figure 6.1). Libya was the first state to become independent in 1951. In September 1969 its monarchical regime was finally removed by an army-based coup headed by Colonel Mu'ammar Qadhafi, the current Libyan ruler. Independence followed for Morocco and Tunisia in 1956, when France decided to abandon its Protectorate regime in each country, in the face of growing indigenous resistance. In Morocco, the original Alawite monarchy, which had been preserved by the French, retained control under King Mohammed V and after 1961 under his son, King Hassan II. In Tunisia, however, the leader of the powerful Neo-Destour independence movement, Habib Bouguiba, dominated Tunisia until November 1987, when he was removed in a palace coup by Zine El-Abidine Ben Ali, the current president of Tunisia. Spain continued to hold on to its colony in the Western Sahara until 1975, when it abandoned it, and it was then annexed by Morocco. The nationalist independence movement, a Sahrawi national liberation movement known as the Polisario Front, was thus left to fight against steadily increasing odds for independence from, first, joint Mauritanian–Moroccan control of the former Spanish colony and then, after

139

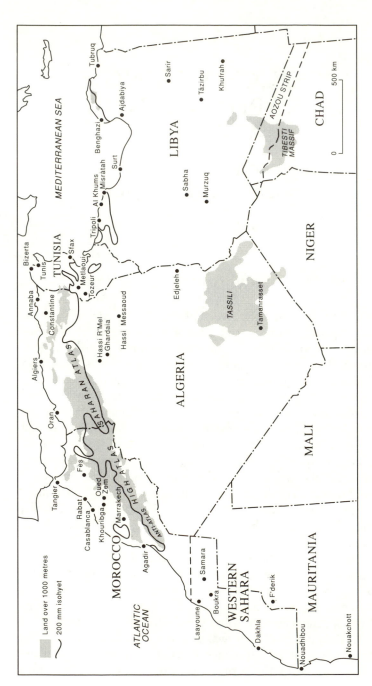

Figure 6.1 North Africa

1979, from Morocco alone (Hodges 1983; Joffé 1987: 21–3). Mauritania had been granted independence within the French Union in 1960, as part of General De Gaulle's post-colonial dissolution of France's African empire. Finally, in 1962, the ferocious Algerian War of Independence was brought to an end, after perhaps as many as 1.5 million Algerian deaths (French official figures are lower), when Algeria, too, was granted independence. The early chaotic socialism of the regime led by Ahmad Ben Bella was replaced by an austere, disciplined regime under Houari Boumedienne, in 1964. After Boumedienne's unexpected death in 1978, his state capitalist ideology was gradually abandoned by his successor, Chadli Ben Jedid.

The last thirty years, therefore, have seen the independent governments of all five countries struggling with the intractable and potentially destabilizing problems of nation-building and economic development. Nation-building has been important because none of the states concerned was inhabited by coherent ethnically homogenous nations which could automatically legitimize the sovereignty of their governments. Economic development has been vital because the colonial period had left all of the new states grievously ill-equipped to satisfy the needs and demands of their populations, and economically dependent on access to European markets in the face of a menacing explosion in population growth.

THE DEMOGRAPHIC TIME BOMB

By far the greatest potential resource – and the greatest threat – for all North African states are their populations, growing at rates of between 2.3 and 3.1 per cent per annum (see Table 6.1). Their

Table 6.1 Population (millions) and growth (1990) in North Africa

Country	growth (%)	Total 1960	1990	Economically active	In education
Algeria	3.1	10.79	25.34	4.49	5.58
Libya	2.3	1.24[b]	4.17	1.06[a]	1.47
Mauritania	2.6	1.00[b]	2.00	0.59	0.18
Morocco	2.9	11.62	25.24	5.74	3.63
Tunisia	2.5	4.16(estd)	8.17	2.14	1.86
Total	3.1		64.92		

Source: Population: *Encyclopaedia Britannica Yearbook*; in education: EIU Country Profiles.
Notes: [a] Includes 136,000 foreign workers.
    [b] 1962.

combined populations already total almost 65 million and will more than double during the next twenty-five years to almost 145 million unless birth rates fall dramatically. The result is that more than half the populations of each country are under 20 years of age – 52 per cent in Morocco in 1988, for example.

Education and employment

The major problem faced by governments is educating these predominantly youthful populations and creating sufficient jobs to satisfy the exploding labour demand. Tunisia, for example, generates 40,000 jobs per annum but there are 70,000 new entrants annually joining the labour market. In Morocco, labour demand grows by over 100,000 annually, while public sector job creation averages around only 70,000 posts each year. The situation is little different in Algeria where 104,000 new public sector posts are required each year, while economic planning is designed to accommodate only 90,000 annually – and the real figure is far lower. Only in Libya is there a shortage of labour, with the result that migrant labour there is officially encouraged.

The result is that unemployment rates are very high. In Morocco, the official unemployment rate was 17 per cent in 1989 (Economist Intelligence Unit 1990–1), although in reality probably closer to 20–5 per cent. In Algeria, unemployment was officially admitted to have reached 22.9 per cent in early 1989 (Economist Intelligence Unit 1990–1). In Tunisia some 13 per cent were recorded as unemployed in 1989, with around half being under 25 years of age.

The problems of employment are paralleled by the demand in North Africa for education. At independence, literacy levels were very low but have risen markedly as a result of intensive educational extension programmes over the past thirty years. These programmes have been particularly directed at primary school students, in order to ensure basic literacy skills. They have also gone a long way to redress the imbalance between male and female education – although less so at higher levels. There is also a growing university-educated stratum in all Maghrebi countries (much of it educated abroad, particularly in Europe) that finds employment opportunities few and far between. In Morocco, for example, there are an estimated 30,000 university and higher education graduates who are unemployed. The situation in Algeria and Tunisia is similar, although, once again, Libya – with its massive oil wealth and small population – is able to provide sufficient employment opportunities. Education is not cheap and between 6 and 8 per cent of Gross Domestic Product (GDP) is dedicated to it every year.

URBAN DRIFT AND MIGRATION

These gross figures for population growth, employment demand and educational need are further complicated by the fact that patterns of settlement morphology are also undergoing radical change. Since the beginning of the colonial period there has been ever-accelerating urban drift throughout North Africa. In the pre-colonial period, urban populations were typically between 10 and 15 per cent of the total population. Within a few years of colonial rule being established, however, this picture began to change and rural populations began to move into town. One major reason was the radical disruptions of traditional agriculture as land was alienated to settler use and tribes were forced to abandon transhumant and extensive agricultural practices. Another was increasing population density as longevity began to increase in the wake of more settled political and administrative conditions, as well as the imposition of major public health measures such as the eradication of malaria from much of the plains area in Morocco. In addition, the traditional rural agricultural worlds of North Africa were soon to be integrated into the monetarized economies being created by the colonial presence and this often reduced the attractiveness of subsistence agriculture. This tendency was intensified by the use of North Africa as a source of cheap agricultural products for the French market and the consequent exposure of North African producers to the vagaries and fluctuations of world agricultural prices (Swearingen 1988).

Since independence, the pressure for urbanization has massively increased, as Table 6.2 makes clear, although it has begun to decline in recent years. The factors responsible for the massive urban drift that has developed over the past thirty years have been substantially the same as existed during the colonial period. They have, however, been made far more acute by the demographic explosion from which all North African countries have suffered and by the development policies that have been put into practice in most of these countries. In every case, traditional agriculture has tended to be neglected in favour of investment in industry, mining and infrastructure.

Table 6.2 Urbanization (%) in North Africa

Country	Urbanization 1965	1984	latest	Average annual growth rate 1965–73	1973–84	latest[a]
Algeria	32	47	49.7	2.5	5.4	4.8
Libya	29	63	67.0	8.9	7.9	7.0
Mauritania	—	—	21.9	—	—	—
Morocco	32	43	45.5	4.0	4.2	3.6
Tunisia	40	54	—	4.1	3.8	—

Sources: Beaumont et al. (1988: 215). For latest figures: EIU, Country Profiles 1990–91, Business International. For Mauritania, Encyclopaedia Britannica (1990: 762–7).
Note: [a] Algeria 1987; Libya 1989; Mauritania 1977; Morocco 1987.

The consequences of such a high level of urban migration have been serious. Quite apart from the demands that migrants make on educational facilities and employment opportunities in the urban setting, many other resources are put under considerable strain. Perhaps the most striking case of this is housing. The influx of migrants has always swamped available urban housing, whether in the old and traditional town centres or in the 'new medinas' created around them. The result has been the growth of unplanned and uncontrolled shanty towns, known in North Africa as *bidonvilles*, from the French word for can. These areas usually lack all or most 'normal' services (roads, sewerage, power and potable water) and are serious potential health hazards. They are also difficult to control and represent a continual potential threat to public order, since their squatter populations are often excluded from the rights and privileges of the formal national economy and governmental administration, such as minimum working conditions.

Indeed, the *bidonvilles* of Tunis, Algiers and Casablanca were at the core of the riots that swept through North Africa during the 1980s (Morocco in June 1981, January 1984 and December 1990; Tunisia in January 1978 and January 1984; and Algeria in October 1988).

Solutions to the migration problem

North African governments have responded to these problems of demography and urban drift in various ways. Attempts have been made to clear squatter settlements and to replace them by low-cost housing schemes, particularly in Algeria. However, new squatter settlements tend to develop. In Morocco and Tunisia, such clearance schemes – often funded by EC and World Bank loans – have been more successful and have been accompanied by government schemes to encourage investment and job creation through the private sector. Often these measures have been accompanied, particularly in Algeria, by the removal of unwanted rural migrants back to their places of origin by the authorities. These clearance operations have been paralleled by rural job creation schemes designed to reduce the attraction of the urban centres.

North African states have also attempted to export their problems. Shortly after independence, governments encouraged labour migration abroad, particularly to Europe where there was a growing shortage of unskilled labour. As a result, a significant North African emigrant population has developed, particularly in France. Other major recipient countries have been Belgium, The Netherlands and Germany. Since the 1970s, however, European states have increasingly discouraged labour migration, both because of growing domestic unemployment and because of the massive problems of social integration that have developed. The advent of the Single European Market in 1993 is expected to mark a definitive end to labour migration from North Africa to Europe. However, the demographic pressure expected in North Africa in the next twenty-five years, together with a significant shortfall in labour supply in Europe which is expected to develop during the same period as the zero-growth European populations age, may well reverse this policy (Chesnais 1990).

To some extent, North African migrants have sought alternative destinations in recent years. Tunisians and Moroccans have sought work in Libya; there has also been some North African labour emigration to the Gulf. Europe, however, has continued to be the destination of choice and North African governments, particularly Tunisia, have sought to persuade the EC to improve the legal status of existing migrants and to allow renewed migration, in order to reduce the pressure on the struggling economies of North Africa which cannot adequately respond to the growing domestic demand for employment. It is also too late (and undesirable) to eliminate the demonstration effect of 'success'. Many migrants who have worked for twenty or so years in Europe return with capital to their native villages and towns, and construct solid houses or open small retail businesses or engage in other aspects of petty commerce or service sector activities. These symbols of what work in Europe can mean will be durable for many years.

NATIONAL ECONOMIC FAILURE

Demographic problems facing the states of North Africa form only part of their current problems, for all of them face serious economic difficulties. At first glance, this is surprising for North Africa has been relatively well endowed with natural resources. Algeria and Libya are major hydrocarbon producers, while Morocco, Tunisia and Algeria are well endowed with arable and pastoral land. Morocco is also the world's largest phosphate exporter and its third largest producer, while Tunisia exports moderate quantities of phosphates and oil (see Figure 6.2). Nor is water the sort of problem that it is in the Middle East, except in Mauritania and Libya. The 200 mm isohyet (the conventional dividing line between arid and non-arid lands) runs east–west along the line of the major mountainous formation in the region, the Atlas mountain complex (see Figure 6.1). To the north of this line lie the major population concentrations and agricultural areas of North Africa. Only Libya and Mauritania lie to its south and even in Libya, the Jefara plain and the Jabal al-Akhdar region enjoy rainfall of around or above 200 mm annually, while in Mauritania the southern border regions benefit from the Senegal river. Tunisia and Algeria north of the Atlas experience a typical Mediterranean climate and benefit from complex river systems debouching into the Mediterranean. Eastern Morocco also enjoys a Mediterranean climate, while the rest of the country benefits from the Atlantic trade winds and the rainfall they bring.

Nevertheless, North African states all face serious economic problems despite these natural advantages. All the countries except Libya suffer from significant and growing foreign debt despite strenuous efforts at debt restraint during the past decade and all of them except Libya have experienced regular current account deficits in recent years, although the measures of economic reform and restructuring taken by all North African states since 1983 have also meant that current account surpluses have been sporadically earned by every one of the states in the region except Mauritania. All of the Maghreb states have also had to engage in economic reforms designed to reduce radically the role of the public sector in the economy, as a result of the economic problems they have faced (for Morocco's case, see Horton 1990). Thus it seems that North African states suffer from economic problems that are common to the developing world. In short, their economic problems are not

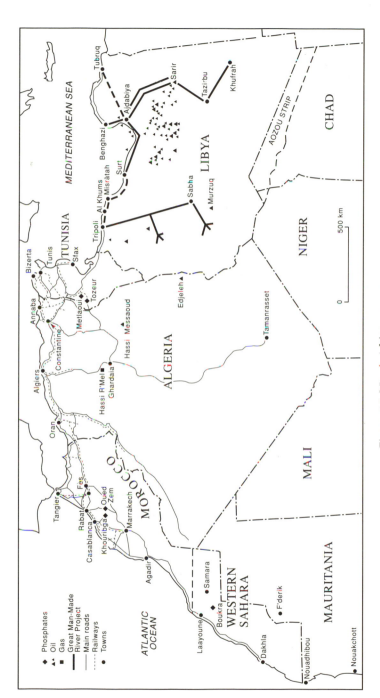

Figure 6.2 North Africa: resources

GEORGE JOFFÉ

just the consequence of individual economic mismanagement but reflect common problems of economic development that arise from the complex and unequal relationships between the developed and developing worlds – the North–South divide.

Those that are oil importers – Tunisia, Morocco and Mauritania – are at the mercy of changes in international oil prices which have progressively increased their energy costs and led them into ever-worsening foreign debt. Those that are agricultural and industrial exporters – Tunisia and Morocco – suffer from discriminatory tariff and non-tariff barriers in their major export market, the European Community. Mineral exporters – Morocco, Tunisia and Mauritania – have suffered from steadily declining or static commodity prices over the past decade. Hydrocarbon exporters – Algeria and Libya – suffer from unpredictable revenue flows which, over the 1980s, have tended to steadily decline as excess OPEC production created a supply surplus (Terzian 1985; Ghanem 1986). At the same time, imports of manufactures to satisfy development and consumer demand have become ever more costly.

Europe: the dominating trade partner

Trade statistics make it clear that the EC is the major trade partner with all the North African states, generating more than half their imports and absorbing up to 60 per cent of their exports. Algeria and Libya play an important part in satisfying European energy demand, supplying 23 per cent of EC oil requirements in 1986 (BP 1990). Algeria also supplies around 25 per cent of EC natural gas requirements each year, a quantity that it would like to double in the 1990s. Algeria and Libya together satisfy some 20 per cent of EC imports of refined petroleum products annually as well. For Morocco, the EC is the major destination for its phosphates, with 54.6 per cent of its raw phosphate exports of 13.7 million tonnes entering Europe in 1986. Tunisia also sent 35.4 per cent of its 1.2 million tonnes of phosphate ore exports to the EC, while Algeria supplied 35.8 per cent of its 0.8 million tonnes of phosphate exports to the EC as well (Sutton 1988).

For Morocco and Tunisia, agricultural goods and textiles have always been dominant elements in the export picture. Agricultural goods form 25 per cent of Morocco's exports and 10 per cent of Tunisia's exports, while textiles form around 17 per cent and 10 per cent of Morocco's. Such exports have been made subject to special

148

EC conditions. These arise from the EC–Maghreb Co-operation Agreements that exist between Brussels and Tunisia, Algeria and Morocco, while Mauritania falls under the EC's Rome Convention arrangements. Only Libya has no such privileged agreement with the EC. It must also be borne in mind that the advent of the Single European Market at the end of 1992 will have a dramatic and largely negative effect on the access of such imports to the EC, as will the unquantified and probably unquantifiable consequence of the dissolution of the communist regimes of Eastern Europe and the USSR. Although industrial goods from countries in the developing world are supposed to have unrestricted access to the EC under the founding Treaty of Rome, after 1992, a new barrier will appear, in that industrial goods from developing countries will have to meet European industrial standards to be permitted entry into the European market – desirable for the European consumer, no doubt, but an added financial burden on manufacturers which will render their goods less competitive.

Agricultural goods suffer from more direct discrimination. Not only do they have to compete against the heavy subsidies paid to European food producers under the EC's Common Agricultural Policy (CAP) – subsidies which may eventually be cut under the GATT's Uraguay Round trade liberalization negotiations – but also they suffer from a specific combined quota and tariff restriction system, known as the Reference Price System. Under this system, a certain amount of produce (mainly citrus and fish productions from Morocco, olive oil from Tunisia and early vegetables from both countries and Egypt) is allowed in duty-free at certain specific periods of the year. Outside these periods and in excess of the permitted quotas, such produce bears a countervailing tariff designed to ensure that their cost to wholesalers equals the basic European internal price for such produce as set by Brussels – the 'reference price'. The simple and unpalatable fact is that North African agricultural produce is no longer needed in the EC, despite the domestic economic and human cost of their exclusion from this traditional market as, because of the productiveness of Spanish and Portuguese agriculture, Europe now produces a surplus in precisely these products which North Africa has traditionally supplied.

Although a new access regime for North African states is to be introduced after 1995, when Spain and Portugal are fully integrated economically into the EC, there are considerable fears in Rabat and Tunis that this will be as restrictive as EC controls on migration are expected to be after 1992. The result is that it will be extremely difficult for countries such as Tunisia and Morocco, which depend on agricultural and industrial trade with Europe, to improve their chronic trade deficits and their predominantly negative current account balances by export-led growth, as required by the IMF and the World Bank. Even non-trade flows – such as earnings from tourism (around $1 billion annually for Morocco and $500 million for Tunisia) and remittances from their migrant workers in Europe (around $1.5 billion annually for Morocco and $250 million for Tunisia) are unlikely to increase significantly after 1992, either because of EC restriction on migration or because of increasing competition in tourism.

State intervention in national economies, and external debt

The economic problems faced by all the North African states really reflect the role of the state in promoting development. Despite the very different economic ideologies in operation in each country, the state has played a major, if not preponderant, role in the economy. Since all the countries based development on five-year planning systems, which were defined and applied by government, this dominant role was inevitable. Development, however, was also heavily dependent on foreign currency, to pay for the imports of goods and services essential to the process. Only oil exporters, such as Libya and Algeria, could anticipate adequate foreign currency revenues, particularly after the dramatic increases in world oil prices in 1973 and 1979. Indeed, Libya still has very low medium and long-term foreign debt – $3 billion in 1988 – as it has successfully financed development from revenues, despite the stagnation and collapse of world oil prices during the 1980s. This has, none the less, involved considerable import compression in the wake of the collapse in world oil prices engineered by Saudi Arabia in 1986. Algeria, however, is massively indebted and now has great difficulty in financing its imports and in servicing its foreign debt. The reason is simply that Algeria has financed major development projects through foreign loans, preserving its oil and gas revenue for current expenditure on imports and for general capital expenditure. This has proved to be a very expensive mistake and Algeria today faces a debt service ratio (annual debt repayment as a percentage of revenues from exports of goods and services) of a staggering 75 per cent. This, combined with declining revenues, as world oil prices fell after

1986, persuaded the Algerian government to seek IMF help. In mid-1989 the IMF granted Algeria a stand-by loan.

The cases of Tunisia and Morocco were somewhat different. Although Tunisia enjoyed some oil income, its role in the general export revenue picture was much reduced. Tunisian leaders have always been acutely sensitive to the dangers of excessive foreign debt and, in 1986, a crisis in foreign currency reserves persuaded them to seek help from the IMF over economic restructuring. The crisis was intensified by drought in 1987 and locust damage thereafter, which forced up Tunisia's food imports at a time when its oil revenues were declining.

Morocco ran into serious debt problems in the late 1970s as a result of over-ambitious development planning in view of its internal financing resources. Expenditure for the 1973–7 plan was revised upwards in 1974, as a result of a fourfold increase in world phosphate prices engineered by the Moroccan state phosphates concern, OCP. Within a year, however, phosphate prices collapsed, while the pressure on Morocco's financial resources was intensified by the costs of the Western Sahara conflict between Morocco and the Sahrawi national liberation movement, the Polisario Front. Instead of abandoning the plan, Morocco increased its foreign debt and, despite stabilization measures in the 1977–80 three-year plan, Morocco had to seek IMF support for debt rescheduling in September 1983. The problems persist, however, and Morocco still suffers from serious problems of foreign debt.

Economic restructuring and the IMF

The IMF has basically required that the three states of Tunisia, Algeria and Morocco undertake radical economic restructuring in return for its help in financing their foreign debts. This restructuring process, which has been supported by additional finance provided by the World Bank, comprises two crucial elements – the reduction of budgetary deficits and the conversion of current account deficits into surpluses. The means chosen for this were trade liberalization and the encouragement of export-orientated industries and other economic activities, together with the reduction of public sector expenditure by cutting consumer subsidies and reducing public sector control of the economy (Joffé 1990a: 30; 1990b: 64; Stevens 1990: 71).

For the domestic economies of North African states, it has been the issues of subsidy removal and of reduction of public sector economic control through privatization that have most obviously affected public opinion. Indeed, subsidy reductions on essential foodstuffs – sugar, cooking oil, flour and bread – were what proved the final catalyst for riots noted above . The effects of privatization have yet to be fully felt, however. These programmes will involve the reduction of state control of the Tunisian and Algerian economies below the current levels of around 40 and 66 per cent of GDP respectively, and of the Moroccan economy from the current level of 20 per cent of GDP in 1988 and 17 per cent of GDP in 1990.

Sectoral problems: modernity and tradition

These changes in economic structure promoted by the IMF cannot, however, help the majority of the populations of North Africa escape from economic underdevelopment in the short or the medium term. Although the private industrial and service sectors may well be stimulated by privatization and trade liberalization, they form only relatively small sectors of the overall economy. Furthermore, economic 'trickle-down' will, in so far as it operates at all, take far too long to alter wealth distribution, given the political imperatives for rapid economic development. In any case, the geographic pattern of economic growth established during the

Table 6.3 Natural resources in North Africa

Country	Area (mn km²)	Agriculture (mn ha)	Oil Reserves (bn bls)	Mineral Reserves (bn tonnes)
Algeria	2.382	42.5 (7.5)[a]	9.2 (3.2)[b]	0.5[c]
Libya	1.760	15.5 (2.1)	22.8 (0.7)	2–3[d]
Mauritania	1.031	39.4 (0.2)	—	2[d]
Morocco	0.711[e]	29.4 (8.5)	—[f] (0.003)	59.5[c]
Tunisia	0.164	7.5 (5.0)	1.8	0.8[c]

Sources: EIU (1990) Country Profiles 1990–91, B.I.; BP (1990; *Encyclopaedia Britannica.*

Notes: [a] Figures in parentheses are for arable land.
[b] Figures in parentheses are for gas reserves in trillion cubic metres.
[c] Phosphates.
[d] Iron ore.
[e] Including the Western Sahara.
[f] Morocco has massive reserves of oil shale, 20 bn tons containing 8 bn tons of oil, which cannot be exploited in viable economic terms at present.

Table 6.4 Origins of GDP(%) by economic sector in North Africa 1987

	Algeria	Libya[a]	Mauritania[b]	Morocco	Tunisia
Primary	31	60	29	28	23
Agriculture	14	4	19	21	14
Mining	17	56	10	7	9
Secondary	32	13	17	33	20
Manufacturing	14	4	8	17	13
Construction	17	} 9	8	6	6
Public utilities	1	}	1	—[c]	1
Tertiary	37	27	44	39	57
Transport/ Communications	5		7	5	5
Trade	16		13	17	23
Financial services		27		15	4
Other services	16		18		13
Government			6	12	12

Source: Encyclopaedia Britannica (1990: 789–91).
Notes: [a] 1985 figures.
 [b] 1984 figures.
 [c] included in mining.

colonial period has been entrenched by the continuing domination of the EC in Maghrebi trade patterns.

Sectors of the economy that fall outside the scope of the type of dynamic modern private sector sought by the IMF and the World Bank are far more important within GDP. These sectors comprise agriculture – in which the traditional subsistence sector far out-weighs the modernized export-orientated sector – general services, particularly petty trade, and natural resource exploitation which tend to still be under state control (see Tables 6.3 and 6.4).

Furthermore, up to 40 per cent of all economic activity is estimated to lie outside the official economy, particularly in Morocco and Western Algeria. Indeed the manufacturing sector is generally less than 15 per cent of GDP, except in Morocco, and privatization, which is predominantly concerned with this sector, cannot there-fore have much direct effect on national economies overall. In the long term, it will be general economic liberalization that has the most significant effect on development and growth, but these are long-term considerations that take little account of the imperatives of improved individual economic standards now.

Ironically enough, attitudes towards development in the past have had much to do with the economic crisis in sectors such as agriculture. One consequence of the colonial period had been to create a dual agricultural system in every North African state except Mauritania. In Morocco, Algeria and Tunisia rural colonization and land expropriation had created an intensively farmed modernized agricultural sector, often dependent on major irrigation schemes and primarily designed to serve the European market. In Libya, the settlement of Italian peasantry under the Fascist regime in the Jabal al-Akhdar and the Jefara plain had a similar effect.

The consequence of this division of land use is that the traditional sector – conventionally set at 80 per cent of the agricultural sector overall – is largely devoted to peasant production of cereals with individual land holdings of between 5 ha and 10 ha, while export crops, such as citrus and olive oil, are confined to the modern sector in which individual land holdings are greater than 50 ha and which generates up to 80 per cent of export revenues. Only Morocco and Mauritania have significant fishery industries, in addition to other agricultural produce.

The only exception to this is Libya where atypically heavy investment in agriculture since 1969 (18 per cent of the total at LD 5.9 billion or $19.9 billion) has created a series of intensive sites deep in the Sahara desert, as well as improving agriculture in the traditional areas of the Jefara plain and the Jabal al-Akhdar. Since 1968, rotary sprinkler irrigation combined with intensive fertilizer application has been successfully used to produce cereals at Kufrah, Sarir and Maknoussa. Yields have not been high and with high production costs and an uncertain aquifer life-span there can be no economic justification for this kind of agriculture. Oil revenues have permitted Libya to engage in economic engineering of this kind. The most spectacular example is the Great Man-Made River project (Allan 1989: 63–72), a massive water pipeline project designed to provide irrigation water to 180,000 ha in the Sirte region and to a further 320,000 ha in the Jabal al-Akhdar and the Jefara plain, as well as satisfying growing urban and industrial demand around Benghazi and Tripoli (see Figure 6.2). The scheme, which will cost over $21 billion if completed, was stimulated by growing salinity in coastal agricultural and industrial regions as a result of overuse of subterranean local water resources.

The fear is, however, that water demand for urban and industrial use will eventually pre-empt supplies and marginalize the agricultural dimension. This is a general tendency and the relative neglect of agriculture in the development process has also had some very serious consequences in terms of food self-sufficiency. Libya, not surprisingly given its extremely harsh environment and population explosion, has not been self-sufficient in food production for several decades and, in 1985, up to 65 per cent of its requirements were imported at a cost equivalent to 19 per cent of Libya's total import bill (Joffé 1988: 50).

It is much less easy to understand why Algeria had to import a similar percentage of its food needs until the mid-1980s and why, despite significant agrarian reform since then, it still imports more than half the food it requires annually. The cause has been the relative neglect of agriculture and its collectivization during the Boumedienne period. Despite an agrarian reform in 1971, the state agricultural sector still controlled one-third of all arable land at the end of the 1980s and state pricing and distribution systems still depressed production. None the less, the encouragement of private enterprise in vegetable, fruit, dairy and poultry production has produced dramatic increases in supply.

The situation in Morocco and Tunisia has been less depressing, although both countries are still net food importers. Tunisia has seen its dependence on food imports grow at 3.8 per cent per year and now produces only 55 per cent of its cereal requirements, for example. Investment in agriculture is being increased in order to reverse this trend. However, the sector still suffers from unpredictable climatic conditions (droughts tend to occur every five to seven years), from unequal land distribution and from inadequate use of inputs, such as fertilizers.

ENVIRONMENTAL CONSEQUENCES

In common with many parts of Africa and the Middle East, North African states are discovering that the combination of demographic explosion and development can carry a heavy environmental and ecological cost. In addition, the very nature of the physical environment in the Maghreb adds to the ecological dangers. The most serious problems are desertification and soil erosion. Soil erosion results mainly from the loss of natural soil cover as marginal land is cleared of vegetation for cultivation. It also occurs as a result of charcoal manufacture, particularly in the Jabala region of Northern Morocco, close to the city of Tangier (Munson 1990: 30–46). The removal of shrub and dwarf palm cover leaves the soil prone to serious erosion through flash-floods in winter and spring. Attempts have been made to restore tree cover, as in the DERRO project in the western Rif mountains of Morocco. In Libya, UNESCO has sponsored an archaeological project to restore balanced land use, based on the modes of cultivation used by Roman settlers in the limes (Barker 1982: 3-8).

Desertification follows a similar pattern and, in some parts of the Northern Sahara, the desert margin has been advancing northwards at up to 30 km annually. It has been encouraged by wind erosion which occurs at certain times of the year and carries massive quantities of red dust up to the coast. In Algeria, attempts have been made during the past two decades to prevent this by a massive tree planting operation along the desert edge. This project, known as 'Le Barrage Vert' which began in 1975, has been the responsibility of the Algerian army. There is now a similar project in Tunisia and Libya has long been trying to fix sand dunes in coastal regions.

There have been two significant and highly publicized cases of environmental threat that have concentrated North African minds on the problem in recent years. First, there was an attempt by a North American company to persuade Moroccan authorities to permit the dumping of industrial waste in the Western Sahara during 1989. The inference that Morocco's environment did not matter and that it could be bought for refuse disposal by the affluent North stirred great resentment, no matter what 'quality' control was assured. Second, oil pollution of Morocco's Atlantic coastline from a damaged Iranian oil tanker threatened to occur during December 1989 and was avoided only by international action. This highlighted the fact that all North African states suffer from maritime pollution, both from general discharge from shipping and, in the case of Algeria and Libya, from spillage and discharge associated with oil terminals and tanker traffic. This is particularly acute in the Gulf of Sirte where four of Libya's five oil terminals are located and around the oil ports of Arzew, Skikda, Algiers and Wahran in Algeria. There is also a problem of industrial effluent and sewage outfalls as in the lagoon of Tunis (Woodford 1990) and around major manufacturing and processing sites, most of which are located on the coast. Serious problems exist around Casablanca and the neighbouring oil terminal of Mohammedia in Morocco.

There have been other, more hidden specific threats as well. One example has been the ecological dangers of prolonged warfare; the locust threat to southern Algeria and Tunisia in 1988 and 1989 was intensified by the war in the Western Sahara which rendered joint regional action over destruction of the swarms in northern Mauritania impossible.

Air pollution too is becoming a serious hazard around towns because of the predominance of a Mediterranean-style climate over most of North Africa – which encourages temperature inversion and, as a result of sunshine, the creation of smog from air-borne dust and chemical pollutants. Water pollution is also becoming a problem, as demand on inadequate water supplies in major cities increases – particularly in Algeria.

ISLAM AND POLITICAL AND CULTURAL MOVEMENTS

The cultural and political environment of North Africa is unusual in the context of the Muslim world for the superficially apparent religious homogeneity of its populations. Apart from small in-digeneous Jewish communities in Tunisia (mainly in urban centres and in Djerba), Algeria and Morocco (in Casablanca, Rabat, Tangier and Marrakesh) and expatriate Christian communities in all five North African states, the population is overwhelmingly Muslim. They form 99.1 per cent of the population in Algeria; 97 per cent in Libya; 99.4 per cent in Mauritania; 98.7 per cent in Morocco; and 99.4 per cent in Tunisia (virtually all Muslims in North Africa are Sunni Muslims).

Islam in North Africa is remarkable in three respects; in the roles played in modern society by religious orders (tariqa-s), Islamic saints (murabit-s) and Islamists. All three factors exist elsewhere in the Muslim world, but the particular combination of them that exists in North Africa has quite specific effects. There is also one aspect of religious practice that is unique to Morocco – the official toleration extended to Judaism. Jewish festivals and holy sites in Morocco receive discreet official support and Jews of Moroccan origin are entitled to Moroccan citizenship even if they do not reside in Morocco.

Most religious orders in North Africa have traditionally bolstered authority and have thus been intimately connected with political and economic power. The Tijjaniya order, for example, dominated the trans-Saharan trade and is today still influential south of the Sahara through the companion Mouriddya order in Senegal. The Sanusi order was the source of the Idrissid monarchy in Libya between 1951 and 1969. Orders such as the Tihamiyya, Rahmaniyya

157

and Qadiriyya still enjoy considerable support. Other orders, dubbed during the colonial period as 'popular orders', have had considerable support, even though they are conventionally considered to be heterodox and contaminated by non-Muslim African traditions. These include the Hamadsha, the Guennaoua and the Issawa.

Murabit-s – persons either assumed to embody ideal human values because of their genealogical links to the Prophet Muhammad or granted such status because of their exemplary lives – also have an important role to play in popular life. Their shrines (zawiya-s or kubba-s) are places of pilgrimage (mawsim/moussem-s) and they are often connected to religious orders of local or regional importance. These pilgrimages are occasions of considerable social importance and involve large numbers – the pilgrimage to Sidi Ali, just to the south of El Jadida in Morocco, involved 200,000 persons in July 1990.

The Islamist movements of North Africa have a far more immediate political significance. They stem in part from the profound Islamic background to daily life and in part from the growing rejection of Western political prescriptions by large groups in North Africa. These movements first became significant in Tunisia at the start of the 1970s, but have also appeared in Morocco and Algeria since the Iranian Revolution in 1979.

In fact, the role of Islam inside political life in North Africa is symptomatic of the profound changes taking place in the region (*Third World Quarterly* 1988a; 1988b). Public hostility to US action during the Kuwait War of 1991 was vehement throughout the Maghreb, and much of the resentment was channelled through Islamic sentiment. All of them depend to a greater or lesser extent on the stabilizing role of informal political systems and demonstrate a preference for political consensus, whether voluntary or enforced, and therefore often wish to as well as have to work with such movements.

Until recently both Tunisia and Algeria operated what were effectively single party systems – the Parti Socialiste-Destourien now called the Rassemblement Constitutionel Démocratique (RCD) in Tunisia, and the Front de Libération Nationale (FLN) in Algeria. Both single parties stemmed from the struggle for independence and had become the vehicles for personal power for the president of each country. In Morocco the monarchy – which evolved from the pre-colonial sultanate with its 900-year long political pedigree – has successfully adapted more modern political structures. Despite attempted *coups d'état* or rebellions in 1971, 1972 and 1973, Morocco has a formal pluralist democratic system which has been in operation since 1975.

In Algeria, however, the dismantling of the state capitalist economic system by President Chadli Ben Jedid after he came to power in 1979 has also brought about the destruction of the FLN's hegemonic control of power. After the riots in October 1988, all controls on other political formations were removed and parties based on religious confessionalism, ethnic identity and political conviction have now emerged. The removal of the army from political life in 1989 means that the Algerian presidency can no longer dictate Algeria's political future and, provided that pluralism can be maintained, Algeria may become the first Middle Eastern or North African state to have a genuinely democratic political system.

Tunisia still has to escape from the trammels of single party politics. Although the president (Zine El Abidine Ben Ali) has attempted to infuse life into the formally pluralistic political system and to create the basis for a national political consensus, the elections for the National Assembly in April 1989 – in which the RCD won all 141 seats – demonstrated his failure. Furthermore, the government has been unable to accept the Islamist Nahda movement into formal political status and faces, as a result, political stagnation. It seems likely that President Ben Ali will find himself forced into reliance on presidential advisers in a political system where democratic representation has become the hegemonic possession of a single political party which lacks genuine popular support. In short, he may have to recreate the 'Carthage Cabal' system for which his predecessor was condemned.

Only in Libya does the baleful perfection of the system of popular democracy enunciated in the *Green Book* and instituted since 1976 exclude a role for political Islam. Under this system, all Libyans are obliged to participate in a system of basic popular congresses which formulate policy on all issues – local, regional, national and international – and which are organized on a local or professional basis.

Day-to-day administration of public services and of the economy is controlled by a series of popular committees which are primarily responsible to the appropriate basic popular congress. Each such congress mandates delegates to express its views on policy at regional congresses and in the general people's congress, the supreme expression of the 'people's authority' (national sovereignty) and which acts as a parliament in other countries. The general people's congress, which usually meets twice a year, annually elects secretaries (the equivalent of ministers) who are responsible for the secretariats and who are grouped together in the general popular committee (cabinet) under the general secretary (premier).

The only problem with this system – which does not permit political parties to operate since they would be divisive within the 'people's authority' – is that it demands consensus and participation from the Libyan population. Not surprisingly, Libyans do not fit these paradigms and, to make the system operate, an activist organization, the revolutionary committee movement, was organized in 1980. It is directly controlled by Colonel Qadhafi (who otherwise has no formal political role in Libya) and is charged with ensuring the safety of the Libyan revolution and of its political structures. This it has achieved with considerable brutality and repressiveness during the 1980s, with the result that many Libyans now live in permanent exile and at least six, rather ineffectual, opposition movements also exist.

In reality, the two-level formal political system described above is sustained by a far more powerful informal political structure. This ensures that all critical positions within the system are controlled by individuals linked to Colonel Qadhafi or his close collaborators by marriage or kinship. The result is that certain tribal groups – the Qadhadhfa, the Maghrara and the Warfalla – have a close control over the economy, the security services and the administration. It also means that other groups, whether confessional or ethnic, are excluded from the political process and that membership of them is sufficient for the individual concerned to be construed to be potentially hostile to the regime. Thus Islamists are excluded and victimized in Libya.

THE ETHNO-LINGUISTIC DIVIDE

These formal political problems are not the only ones which bedevil North African politicians. They also face a dual linguistic problem which is, in part, a consequence of their common colonial experience. It is a problem really confined to the Maghreb states of Tunisia, Algeria and Morocco, although Mauritania also faces ethnic difficulties and Libya has experienced ethnic tensions in recent years.

As a result of the French colonial presence, the three Maghrebi

160

states have long depended on French as a linguistic vehicle for commerce and administration. Maghrebi governments, however, have argued that Arabic should be the vehicle of official linguistic expression. As a result, Arabization campaigns have been undertaken in all three states (and is now also underway in Mauritania). There have been serious problems, however. Many North Africans consider modern literary Arabic to be a Middle Eastern language and resent its imposition for political purposes – despite Classical Arabic's religious significance as the language of the Qur'an and their association with the Middle East over developing world issues.

The problem is complicated by the fact that opposition is not only based on the practical difficulties involved in Arabization – lack of suitable teachers, isolation from Europe or difficulties of administration and commercial contact. It is also based on the fact that significant minorities in Algeria and Morocco (together with very small minorities in Tunisia and Libya) do not speak Arabic as a first language or do not speak it at all. They use one of several Berber languages instead.

North African governments have taken varying stances over this issue. In Libya, the issue does not arise, since Arabic has been the official language for many years. In Tunisia, the government has refused to abandon the use of French for professional and commercial purposes, although Arabic is the official language. The same has been true in Morocco and there has also been a wary toleration of Berber languages as well. In Algeria, however, the government has been far more insistent on Arabization, although it has so far been unsuccessful in applying it (Zartman 1985: 24–35). The language issue has severe political overtones for Berber groups often felt excluded from the political process. In April 1980, there were serious riots in Kabylia over the language issue and the recent Arabization law passed in Algeria seems likely to set off the confrontation again.

It is important to realize that the Arab–Berber issue is not really ethnic in nature. The ethnic makeup of the populations of North Africa is diverse yet in a sense substantially homogeneous, being derived from both Berber and Arab origins over the past millennium. This is not the case in Mauritania, however, where there are substantial non-Arab-speaking black minorities (350,000 out of the 1.94 million strong population in 1989) and there have been serious racial confrontations during 1989 and 1990 between these two groups. These racial tensions are likely to persist.

GEORGE JOFFÉ
THE GEOGRAPHY OF IGNORANCE

It is clear from what has been said above that the outlook for North Africa is not very promising. In addition to the massive demographic problems that face each state, the opportunities for economic development seem increasingly restricted. Furthermore, the international political environment offers little encouragement. In many ways, it appears that the outlook for North Africa today is more clouded than it seemed to be at independence. The need for effective research into the problems facing the region is, therefore, more acute than ever.

Certain types of problems spring immediately to mind, particularly those connected with demography and development. We do not properly understand the pressures causing urban drift, nor do we understand the workings of the informal economy that mitigates many of the worst consequences of the demographic explosion. There are also a series of acute questions connected with water and land use in order to sustain an agricultural work-force and yet to maximize food production. Little work has been done on appropriate models of agrarian reform, particularly in connection with major irrigation schemes of the kind now being introduced in the Maghreb states. Nor has anything other than scant attention been paid to the ecological and environmental implications of economic development.

Attention must also be directed towards the issue of economic restructuring. It is still not clear that privatization is the best method of dynamizing the private manufacturing sector in countries such as those in North Africa. There is also the problem of debt restructuring and repayment. Tunisian politicians have proposed, for example, that a new development bank for the Maghreb should be created. This would receive debt repayments made by North African states and would use such funds as capital for development investment back into the region. Such ideas may not be attractive to the original lenders, but they may well ease the constant crisis over adequate investment funds for the region, while also making regional politicians responsible for the development process. Attention must also be paid to the implications of North Africa's continuing relationship with Europe – there is hope for greater political co-operation and economic integration with the recent formation (February 1989) of the Union Maghreb Arabe.

There are, in short, a mass of issues that require investigation. The problem is that the means by which this can be done – trained personnel and adequate funding – are still lacking. To a large extent, this requires a commitment from European states and a realization by European politicians that they are also intimately involved in and affected by the development process. To be fair, some European states do appreciate this. France has had a long-standing interest in North Africa overall. The German Federal Republic has also devoted considerable resources to geographical investigation of Morocco. Italian geographers have been involved in Tunisia and Libya, while Spanish specialists have begun to take an interest in Morocco as well. The Anglo-Saxon world, however, lags far behind. It is hoped that the next twenty-five years will see this position reversed.

REFERENCES

Allan, J.A. (1981) *Libya: The Experience of Oil*, London: Croom Helm.

Allan, J.A. (1989) 'Natural resources: not so natural for ease of development', in Allan, J.A., McLachlan, K.S. and Buru, M.M., *Libya: State and Region – A Study of Regional Evolution*, London: SOAS-CNMES.

Barker, G. (1982) 'Natural resource use: lessons from the past', in Allan, J.A. (ed.) *Libya since Independence*, London: Croom Helm.

Beaumont, P., Blake, G.H. and Wagstaff, J.M. (1988) *The Middle East: A Geographical Study*, London: Fulton.

Bennoune, M. (1988) *The Making of Contemporary Algeria, 1830–1987* Cambridge: Cambridge University Press.

BP (1990) *Statistical Review of World Energy*, London: British Petroleum.

Buren L. van (1988) 'Mauritania', in *Arab Agriculture 1987 Yearbook*, Bahrein and London: Falcon.

Chesnais, J-l. (1990) 'Africa's population explosion spurs exodus of youth to Europe', *Guardian*, 29 December, London.

Economist Intelligence Unit (1984) *Annual Regional Review: Middle East and North Africa*, London: EIU-B1.

Economist Intelligence Unit (1990–91) *The Middle East: Country Profiles.* London: EIU.

Encylopaedia Britannica (1990) *Book of the Year*, Chicago.

Ghanem, S.H. (1986) *OPEC: The Rise and Fall of an Exclusive Club*, London: KPI.

Hodges, A. (1983) *Western Sahara: Roots of a Desert War*, London: Croom Helm.

Hopkins, M. (1989) *Tunisia to 1993*, London: Economist Intelligence Unit-BI.

Horton, B. (1990) 'Morocco: analysis and reform of economic policy', *Development Case Series, No 4*, EDI, Washington DC: World Bank.

International Monetary Fund (1990) *The World Economic Outlook*, Washington DC: IMF.

Joffé, E.G.H. (1987) 'The International Court of Justice and the Western Sahara Dispute', in Lawless, R. and Monahan, L. (eds) *War and Refugees: The Western Sahara Conflict*, London: Pinter.

Joffé, E.G.H. (1988) 'Libya', in *Arab Agriculture 1987 Yearbook*, Bahrein and London: Falcon.

Joffé, E.G.H. (1990a) 'Privatisation: the Moroccan experience', in *Privatisation and Economic Reform in Morocco*, Chamber Economic Reports no. 29, ABCC, London.

Joffé, E.G.H. (1990b) 'Privatisation and decentralisation in the Arab World', *JIME Review* 8.

Knapp, W. (1977) *North West Africa: A Political and Economic Survey*, Oxford: Oxford University Press.

Lawless, R. and Findlay, A. (eds) (1984) *North Africa*, London: Croom Helm.

Morad, M. (1990) 'Towards a protection policy for the environment in the Arab World', *Arab Affairs* 11.

Munson, H. (1990) 'Slash and burn cultivation and charcoal making, emigration fromn the highlands of Northwest Morocco', in Murdock, M.S. and Horowitz, M.M. (eds) (1990) *Anthropology and Development in North Africa and the Middle East*, Boulders Col.: Westview.

Royaume du Maroc (1988) *Situation Démographique régionale au Maroc: analyses comparatives*, Rabat: CERD (Centre d'études et recherches démographiques).

Stevens, P. (1990) 'Privatisation in the Middle East and North Africa', *Arab Affairs* 10.

Sutton, M. (1988) *Morocco to 1992: Growth Against the Odds*, London: Economist Intelligence Unit.

Swearingen, W. (1988) *Moroccan Mirages*, London: I.B. Tauris.

Terzian, P. (1985) *OPEC: The Inside Story*, London: Zed Books.

Third World Quarterly (1988a) 'Succession in the South', 10 (1), London: Third World Foundation.

Third World Quarterly (1988b) 'Islam and politics', 10 (2), London: Third World Foundation.

Wright, J. (1982) *Libya: A Modern History*, London: Croom Helm.

Woodford, J.S. (1990) *The City of Tunis*, Wisbech: Menas.

Zartman, I.W. (1985) 'Political dynamics of the Maghreb', in Barakat, H. (1985) *Contemporary North Africa: Issues of Development and Integration*, London: Croom Helm.

7

THE CHANGING GEOGRAPHY OF THE LOWER NILE

Egypt and Sudan as riparian states

J. Anthony Allan

INTRODUCTION: THE BEGINNINGS OF INDEPENDENT DEVELOPMENT

Egypt and Sudan began their political and economic development independent of imperial Britain on the basis of different resource endowments, with very different infrastructural developments and with very different relationships with the occupying power, Britain (see Figure 7.1). Egypt became nominally independent in 1922 after forty unhappy years of British occupation (itself following some years after the heavy French involvement at the beginning of the nineteenth century). The frustration of the Egyptian people with the next three decades of dependent monarchy exploded in the July 1952 Revolution when the ineffective King Farouk was toppled and Colonel Nasser, later President, came to power. Sudan's independence occurred in a much more clear-cut fashion in 1956 as power was handed over directly by the British colonial government to an independent government of the Sudan during the first phase of withdrawal by Britain from its African colonies in the 1950s. The British presence in the Sudan, though comprehensive, had been very thinly spread and the development of the country was at a very preliminary stage, so that by the 1960s there were only a few miles of tarmac road, very limited rail communications and one major agricultural scheme, the Gezira. The development of water resources by controlling dams had begun but the volume of water commanded within the Sudan was less than 30 per cent of that which would be at its disposal according to the Nile Waters Agreement of 1959 between Egypt and the Sudan.

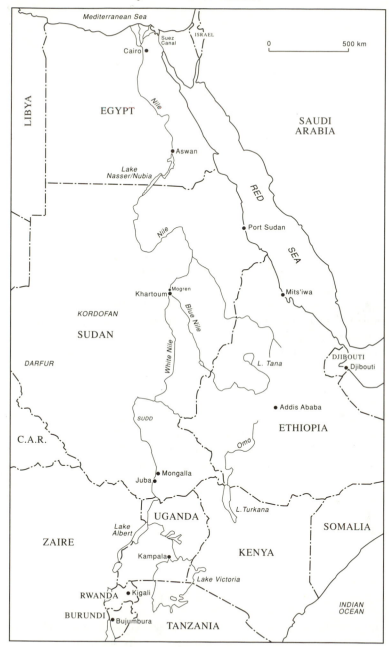

Figure 7.1 The countries of the Nile Basin

166

EGYPT IN THE MID-1960s

For over a century the Suez Canal had been a prime element in the economy of Egypt although, unhappily, a negative one. Despite the Egyptian inputs of labour and partial finance little benefit came to Egypt because control of the Suez Canal Company and its revenues were ceded to a European company soon after its completion. The nationalization of the Suez Canal Company by Colonel Nasser in 1956 represented for Egyptians the belated recovery of an asset whose benefits had been denied to the Egyptian people for almost a century. Nasser was able to demonstrate that the Canal was no longer a legitimate imperial interest of the declining European powers when he confronted the Anglo-French attempt to take the Canal back into international control by the miltary invasion of 1956. Happily the United States confirmed Egypt's position during the 1956 Suez Crisis by denying support to the extraordinary Anglo-French-Israeli collusion.

The Suez Crisis marked a turning point in Egyptian internal and external affairs and it set Egypt on the course which determined its political, economic and institutional shape and international posture in the mid-1960s, the beginning of the period being reviewed. By 1965 Egypt was a socialist state where the central direction of the economy and of people's everyday lives was all pervasive. Priorities in agriculture, industry and civil affairs were set by massively expanded civil service ministries. Nasser focused on issues of essential national importance and brought resources to bear upon them. For example Egypt's need for a secure supply of water for agriculture and power was addressed as the prime concern. Basin wide approaches to storing and managing Nile waters which had preoccupied British engineers and officials (Hurst et al. 1966) were abandoned in favour of an agreement with Sudan enabling a new dam to be built at Aswan with the capacity to store three times Egypt's anticipated level of annual utilization. The USSR also contributed 25 per cent of the capital costs of the High Dam after arrangements with the World Bank fell through in 1956 (Little 1965: 47). The dam began to impound the river from May 1964 and by 1967 the annual Nile flood, which had inundated Egypt's fields every September since before history, ceased.

That the Soviet leader Krushchev participated in the ceremony, closing the sluices to start impounding the waters at the dam at Aswan in 1964, was of great symbolic and real significance. As early as 1955 an arms deal signalled that the Soviet Union was wanting to assert itself in the Middle East, and also that Nasser wanted to disentangle Egypt from its traditional dependence on Western powers. The leadership of the Arab World assumed by Colonel Nasser, especially after the Suez Crisis, had increased the importance of Egypt to the superpowers, and Nasser's influence was greatly enhanced by his clever exploitation of 'Cold War' politics in playing off one superpower against the other.

At the beginning of the review period Egypt was well advanced on an experiment to develop its natural, human and institutional resources with imagination and confidence. There was predictable external criticism of Egypt's attempt to grapple with its development challenges. Everything was criticized, its defence priorities, foreign excursions and especially Egypt's development policies, with a great deal of negative comment about the viability of the High Dam project. These criticisms came mainly from those who had lost influence in the region, Britain and France, but in due course the United States and inevitably Israel joined in (Al Ahram 1975). Elsewhere, however, there was a general sense of international goodwill towards Egypt in its ambitious endeavours.

The challenges were immense. Population was rising rapidly and the irrigated area per person was falling. There was an urgent need to expand agricultural production and productivity as well as to cope with the problems of rapid urbanization involving unprecedented levels of construction and industrial development as well as an expansion of the country's material and social infrastructure. These were the same challenges facing many countries throughout Asia and Africa and the Nasser years involved an interesting comparative socialist experiment in attempting to accelerate the pace of economic and political change. The comparisons would be made with other experiments in other circumstances, with for example the countries of sub-Saharan Africa, with other populous countries such as India and China, and with the countries of what twenty years later was to be called the Pacific Rim. In the event the progress of Egypt has been remarkable in that it has survived as a stable political entity despite the pressures of rapid population growth and limited natural resources. In comparison with other African countries Egypt has performed well in that the vast majority of its people are much better off by 1990 than were the peoples of sub-Saharan Africa. At the same time it is clear that

Egypt has not been an industrial success and advances in industrial development have been modest.

Only part of the relative success of Egyptian development can be attributed to Nasser, who died in 1970. The 1970s proved to be years when more market-orientated systems were reintroduced but the centralized style of the Nasser years survived him in that the bureaucracy which he created proved to be both durable and expansive but not easily dismantled. Nasser's successors, President Sadat from 1970 and President Mubarak from 1981, have adapted to the economic deficits by major shifts in foreign policy, for example by the abandonment of the potentially rich niche afforded by the ability to exploit 'Cold War' tension and very importantly through the extremely awkward accommodation with Israel. These adjustments have been of great significance at home in economic terms because they have enabled a stream of assistance to come from the United States which has been crucial in maintaining Egypt's social and political stability. Defence spending has also been reduced relatively. Above all, however, they have enabled the substitution of US dollars for deficits in water and in agriculture, and in numerous ways they have facilitated the closure of Egypt's growing food gap, while enabling the nation's economy and infrastructure to continue to develop.

THE SUDAN IN THE MID-1960s

The Sudan became the largest (in area) independent country in Africa in 1956 and its government assumed mighty problems as, according to any measures of economic performance, it was one of the poorest countries in the world. Sudan was even more obviously a deficit economy than Egypt, although there was and remains dispute about the ultimate economic potential of the country. The debate was to continue through the 1960s and the 1970s and even into the early 1980s, as some experts judged the resources of the Sudan to be potentially rich and the country to be far from being an inevitable debtor but rather the potential source of agricultural products for the whole of the Middle East and beyond (Hopper 1976: 121). Experience has shown such heady optimism to be misplaced, but at the same time the past twenty-five years have not been favourable for the testing of the hypothesis. Both nature in the form of rainfall deficits (Hulme 1990) and more important the political instability of the south (Howell 1973; Collins 1990) have cruelly impeded the development of natural resources in agriculture.

The northern half of the Sudan is desert and the rest of the country is subject to seasonal rains and because of the nature of the soils of the south the rains make the land difficult to work and the terrain impassable for months at a time. The approach of the British colonial administration had been to develop water resources for irrigation in major projects such as in the Gezira but only in a way which would not significantly impair the interests of its down-stream 'colony', Egypt. The approach to development in the colonial period was therefore tentative, partly because of the Sudan's difficult resource endowment which made surpluses for reinvestment difficult to achieve and partly because the decision-makers in London had limited vision and limited resources at their disposal. An enclave economy was created in the midst of a country otherwise involved in subsistence agriculture and animal rearing. The verdict on colonial agricultural development varies from favour-able (Gaitskell 1951), to critical (Waterbury 1979) to extremely critical (Barnett 1975).

Sudan's leaders at independence attempted to grapple with the developmental challenge in the rural sector. In Sudan, in contrast to Egypt, the majority of cultivated land is rain-fed. Of the 11 million feddans (4.6 million hectares) of cultivated land 85 per cent are rain-fed. Only a small proportion of the arable area was cultivated by mechanized means in the 1960s and there has been a significant expansion of mechanized farming especially in the 1970s. By the mid-1980s approaching one-third of rain-fed farming was mechan-ized. The easiest tracts to farm have always been the clay plains in the centre of the country wherever these could be commanded by Nile water. The soils of Kordofan and Darfur in the west have light, sandy soils capable of producing sesame, groundnuts and water-melons and other vegetables. The soils of the Equatorial provinces in the south have little potential for agriculture.

Although irrigated farming comprised only 15 per cent of the national cultivated area of the Sudan by the mid-1960s it already accounted for the majority of the nation's crop production. Irrigated farming grew steadily from the beginning of the twentieth century as the result of the construction of dams on the Blue Nile and its tributaries and on the White Nile (see Table 7.1).

Sudan had already established a significant irrigated sector at the beginning of the review period and its leaders had the intent to expand this substantially over the next quarter of a century, as well as to extend the area of rain-fed cultivation.

Table 7.1   Water storage in Sudan and Egypt by 1966

Reservoir	River	Year	Live storage (bn m³)	Annual evaporation (bn m³)
Sennar	Blue Nile	1925	0.6	0.3
Jebel Aulia	White Nile	1937	3.5	2.5
Khasm el Girba	Atbara	1964	1.3	0.1
Roseires	Blue Nile	1966	2.7	0.4
Lake Nasser/Nubia	Nile	1963	107.0	10.0
Total				13.3

Source: Chesworth (1990: 47)

THE DEVELOPMENT OF LAND AND WATER IN EGYPT AND THE SUDAN MID-1960S TO 1990

Egypt

The urgency to extend the cultivated area in Egypt in the nineteenth century had been driven mainly by commercial considerations. Egypt needed revenues and the agricultural sector could provide these by increasing the production and export of cotton. By the early years of the century, by which time the population had increased to over 10 million, the need to increase food production also became evident. The response of the Egyptian agricultural sector is shown in Table 7.2. Except in the 1930s there has been a steady increase in the cultivated area. The performance in the 1960s is possibly understated by the data in Table 7.2 and it is not yet possible to establish a final figure for the extent of sustainable irrigated farming achieved in the 1960s following the increase of water availability following the commissioning of the Aswan Dam. The viability of the land reclamation ventures of the 1980s has not yet been finally established. It is also likely that the total cultivated area was by 1990 significantly higher than the 6 million feddans (2.5 million hectares) estimated for 1986. Precise land use and cropping data are difficult to establish in Egypt because not only is the irrigated area being constantly extended but also irrigated land is being lost to construction.

In order to supply the Egyptian agricultural sector with water and at the same time provide power for all sectors of the economy Nile water has to be carefully managed at the Aswan Dam. Egypt agreed with Sudan in 1959 that it would take up to 55.5 billion cubic

171

Table 7.2 Population and cultivated and cropped land in Egypt

	Estimated population (millions)	Cultivable area ('000 ha)	Cropped area ('000 ha)	Cropping intensity
1821	2.5 to 4.2	1,273	1,273	100
1907	11.19	2,391	3,167	141
1937	15.92	2,215	3,462	156
1966	30.08	2,502	4,337	173
1975	37.00	2,377	4,462	188
1986	49.70	6,000	11,400	190

Sources: Waterbury (1979) for population and cropping data to 1986, and Chesworth (1990: 45).

Note: Official data on irrigated area are regarded as unreliable. A survey based on Ministry of Agriculture studies supplemented by data from satellite imagery indicate that the irrigated area in 1990 was close to 2,919,000 hectares (Ministry of Agriculture 1990).

Table 7.3 Storage and releases (bn m^3) at the Aswan High Dam 1970–89

Year	Storage end July[a]	Release	Evaporation and seepage losses
1971		55.9	10.7
1976	108.4	54.7	15.0
1980	103.1	56.7	12.8
1986	(53.7)[b]	55.5	5.7
1987		<55.5	—
1988	(41.4)	55.5	—
1989	(75.3)	55.5	—

Source: Permanent Joint Technical Commission (1970–87); storage data from Stoner (1990: 84).

Notes: [a] July is lowest level annually.
[b] Figures in parentheses are all below secure levels of storage.

metres per year, and it was implied that any water from the 18.5 billion cubic metres per year of Sudan's entitlement which was not used in any year by Sudan would also be available to Egypt. Storage in, and releases from, the Lake Nasser/Nubia have been as shown in Table 7.3.

Egypt has been able to utilize more than its 1959 Nile Waters Agreement level of 55.5 billion cubic metres in almost all years after 1971. By 1980 the level of Lake Nasser/Nubia was so high that damaging overflows were envisaged. In the event the overflow sluices were never required because the flood levels recorded from 1980 onwards were so low that the accumulated deficit in annual flows meant that land had to be withdrawn from cultivation in 1987 because of a shortage of water and major reductions in cropped area were envisaged for the 1988–9 season. In the event the 1988 flood was one of the highest on record and Lake Nasser recovered to ensure water security for 1988–9 and the following three seasons.

Sudan

Just before the start of the review period Sudan negotiated a 25 per cent share of the estimated long-term flow of the Nile, in other words 18.5 billion cubic metres of 84 billion cubic metres. While Sudan has sufficient irrigable land to make use of this volume of water and it does intend to use it, it has not proved to be possible to control and utilize its share of water in the period since the mid-1960s (Abdel Salam 1976; Wynn 1971). There was a steady increase in the area irrigated in the 1960s; in the 1970s this trend slowed and was followed by a decline in irrigated area in the late 1980s. Sudan has so far never used more than 13.1 billion of the 18.5 billion cubic metres to which it is entitled (Permanent Joint Technical Commission 1970–87).

The irrigation schemes in place in the mid-1960s were extended and new schemes established. Table 7.4 indicates the position achieved by the mid-1980s as well as the major plans which exist to increase the utilization of Nile water. One of the major ventures of the 1970s and the 1980s was the sugar production and refining project proposed by the British-based Lonrho corporation. The new government of Nimeiri had inherited a land management system which nominated most of the land of Sudan as government property and Nimeiri also instituted sweeping nationalization of other institutions. By 1975 the Kenana Sugar Scheme was beginning to show progress. By the end of the decade the enterprise was a significant element of the Sudan's agricultural economy.

THE UNPREDICTABLE NILES: SOME EXTRAORDINARY GEOGRAPHICAL EVENTS IN THE NILE BASIN 1960–90

By far the most interesting geographical phenomenon of the review period in the north-east of Africa occurred in the Nile itself. As a result of two decades of environmental observation in the Upper Nile catchments, the expectations of informed people in Egypt and the Sudan moved from being optimistic about the resources of the Nile to being painfully aware that the Nile will not be able to provide sufficient water for existing agricultural sectors, never mind for the ambitious schemes envisaged to reclaim land in the future. And all of this before the possible reductions of flow which could follow the development of water for agriculture in Ethiopia and the other riparian upstream interests.

Table 7.4 The Nile Waters Agreements of 1929 and 1959

	1929 Agreement (billion m³)	*1959 Agreement (billion m³)*
Egypt's share	48.0	55.5
Sudan's share	4.0	18.5
Unallocated	32.0	0.0
Storage losses	0.0	10.0
Total	84.0	84.0

Other 1929 Provisions
Egypt has right
(i) to monitor flows in Sudan
(ii) to undertake projects without consent of upper riparians
(iii) to veto any construction projects against her interests
(iv) to receive the entire timely flow of water (20 January to 15 July)

Other 1959 Provisions
(i) A Permanent Joint Technical Commission to oversee the Agreement
(ii) to co-ordinate a joint Egyptian–Sudanese position in negotiation with other riparians and to build water conservation projects
(iii) equal sharing of any natural increase in water yield, and equal sharing of any natural loss
(iv) equal sharing of costs and benefits from engineering works in Sudanese upper Nile, e.g. Jobglei and Sudd schemes.

In the mid-1960s Egypt and the Sudan were as dependent on the Nile as they had been for millennia. Its seasonal floods – seemingly predictable – enabled productive agricultural economies within a zone which, outside the narrow strip of inundation, was barren desert. But change was imminent and the thirty years between 1960 and 1990 provided an extraordinary example of how Nature can be unpredictable and can inflict changes in the environment which are at best confusing and at worst extremely inhibiting of existing and planned economic activity. Since the mid-1960s the countries of the Upper Nile Basin have experienced short-term changes in climate and consequently the flow of the Nile to Sudan and Egypt has been markedly different from its flow during the previous century (Hulme 1990). While the Blue Nile was performing according to long-term average flow rates in the 1960s, by the early 1970s, after five years of low rainfall in Ethiopia and in the southern Sudan, there were signs that the assumptions underlying the 1959 Nile Waters Agreement between Egypt and the Sudan and the design of

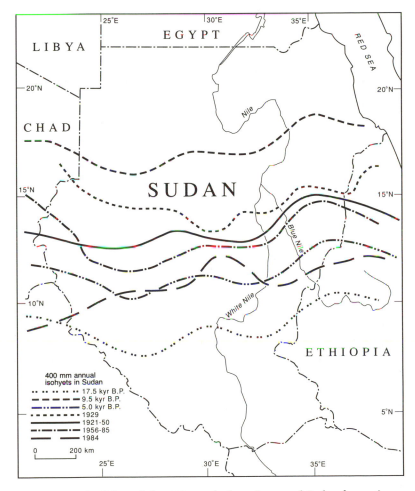

Figure 7.2 Position of the 400 mm isohyet in central Sudan for various twentieth-century years and periods, and for three Holocene centuries
Sources: Wickens (1982) for Holocene estimates; Hulme (1990) for twentieth-century data.

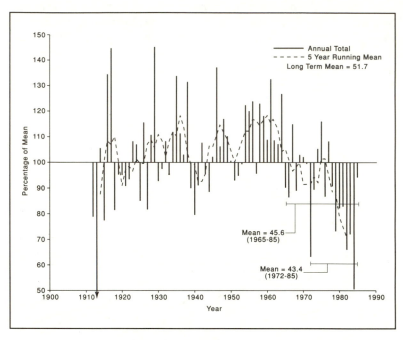

Figure 7.3 Depatures from the mean of naturalized annual Blue Nile flow
at Khartoum 1912–85
Source: Evans (1990: 37).

the High Dam at Aswan were suspect. The 1980s proved to be even
more difficult than the 1970s because the rains in Ethiopia and the
Sudan were 20 per cent below the long-term average not only
causing the famines that were reported world-wide, but also causing
water shortages downstream that received little attention in the
world's press. But those responsible for planning the use and
development of water had to begin to think very carefully indeed
about water management strategies.

The most interesting aspect of this environmental puzzle was,
however, the performance of the White Nile. The White Nile drains
six countries in East Africa and has some large natural storage lakes,
the largest of which is Lake Victoria. The rains in Kenya were so
very much above average in 1962 that the level of Lake Victoria rose
by over two metres and since the lake is the second largest inland
water body in the world, the rise in level represented a huge increase
in storage and enabled a very significant augmentation in the flow of

176

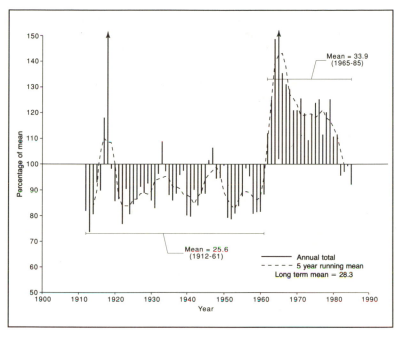

Figure 7.4 Departures from the mean of naturalized annual White Nile
flow at Mogren 1912–85
Source: Evans (1990: 37)

the White Nile for the next twenty years. The White Nile loses half
its volume as it traverses the Sudd (swamplands) of the southern
Sudan, but there was still sufficient water reaching Khartoum to
mask, until the early 1980s, the deficit in the Blue Nile tributaries.
The late 1970s and the early 1980s saw the beginning of an even
more serious period of below average Blue Nile flow than that of
the early 1970s and the combination of lower, but normal, White
Nile flows with the accumulating deficit from Ethiopian tributaries
created the crisis of 1988. The environmental problems were trouble-
some in themselves in terms of environmental management and the
performance of agriculture but their impact was exacerbated in Egypt
by the relentless increase in population (see Figures 7.2 to 7.6).

The period of review started optimistically as it was assumed that
the major engineering works at Aswan would secure Egypt's water
for fifty years at least. But by 1988 it had become evident that the
dam was not adequate to provide permanent protection. Discharges

from Lake Nasser had to be reduced because there was insufficient live storage in the lake to meet the agricultural demands for water. The experience of the 1980s cast a dark shadow over the resource future of Egypt. The 1980s have witnessed a revolution in expectations concerning the availability of water in the Nile system. The 1990s will witness an equivalent revolution in the behaviour of the two major users of Nile water, Egypt and Sudan. The Egyptian–Sudanese 1959 Nile Waters Agreement was never signed by Ethiopia from which comes over 80 per cent of the long-term average flow in the system. In addition President Nyerere of Tanzania explicitly repudiated the 1959 Agreement, and this approach has been adopted, mainly tacitly, by the six East African White Nile countries (Okidi 1990). It is clear that Egypt as the major user will have to identify policies which indicate a willingness to reciprocate in return for a secure supply of water. The clearest potential political link is between Egypt and Ethiopia but this is only in the short term. If the Sudan could settle its debilitating conflict in the South then it would be able to complete its water-using projects and it would inevitably use more water as well as contribute to the hydropolitics by asserting its right to do so.

The lesson which has been spelled out most clearly in the past twenty-five years in the Nile Basin is that political and social factors are the most influential in determining economic and organizational outcomes. The economic predicaments and futures of Egypt and of the Sudan have certainly been affected by the extraordinary environmental events of the recent past, but it is the demographic factor, a social factor, which has determined the scale of Egypt's food gap. In Sudan and Ethiopia politics have had a very detrimental effect on the development of water. The lethal impact of the disruptive military conflicts in both Ethiopia and the southern Sudan have determined the pace of development and generally halted it. In the case of Ethiopia military conflict has prevented significant progress in developing Nile water and in the southern Sudan the people of the South have taken the extreme political action of vetoing the Jonglei Canal Scheme which would have increased by 5 per cent the usable water in the system. The political impacts have both reduced the use of water in Ethiopia and the Sudan and have therefore increased its availability in one sense for downstream Egypt, while simultaneously preventing the conservation measures which would have increased the volume of water by such schemes as the Jonglei projects.

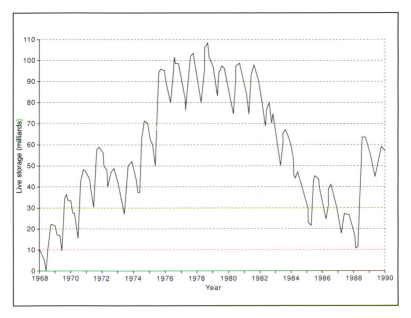

Figure 7.5 Lake Nasser live storage
Source: Chesworth (1990: 57)

At the same time political adjustments have had extremely important positive impacts, for example the changes in Egyptian foreign policy towards the United States have ensured that food supplies have been assured throughout the period of the review despite the sharp widening of the food gap since the early 1970s. The hydropolitics of the Nile basin are evolving rapidly and Egypt's expectations that evidence of long term prior use would ensure future supplies of Nile water are being modified as the interests of the upstream states are more clearly articulated (Republic of the Sudan 1955; Abdulla 1971; Okidi 1990; Abate 1990).

Another inescapable international political aspect of the future development of Upper Nile water is the need for international finance. All the upstream states, and especially Ethiopia, the major source of water, face medium and long-term severe economic deficits and they could not raise the investment to initiate the major structures to control the rivers. The role of the United States will be crucial as will be those of major international agencies such as the World Bank and major economic entities with surpluses such as

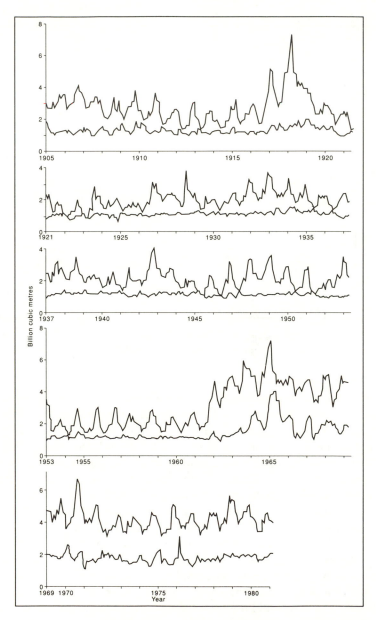

Figure 7.6   Monthly Mongalla inflows and outflows from the Sudd
1905–80 (in billion cubic metres)
Source: Sutcliffe and Parkes (1987)

180

Japan. In the absence of 'Cold War' tension the opportunities to bid up the contributions of the major donors and investors in exchange for strategic influence are for the moment limited. Once international hydropolitics have mobilized the necessary investment resources the pace at which the development of the Nile will proceed will rest with the peoples of the Upper Blue Nile catchments and the extent to which they can successfully assure their financial sponsors that adequate reciprocal arrangements have been made with respect to the major downstream users.

ECONOMIC PERFORMANCE

The first fifteen years of the period under review were relatively favourable for Egypt and the Sudan. Economic indicators were satisfactory and in the case of Egypt promised the possibility of sustained growth. Egypt's better fortunes arose because of its attention to water management at the beginning of the 1960s through the construction of the High Dam at Aswan and during the same decade the equally vigorous development of the country's infrastructure and industry. Egypt was able to survive during the 1980s because of a change of posture towards the West and especially by the decision after the 1973 war to disentangle itself from confrontation with Israel. The policy was implemented by President Sadat and culminated in the Camp David Accord of 1977. Since the actual strategic importance of Egypt was unchanged by this shift in posture the move brought the economic patronage of the United States, and while this relationship has many disadvantages for Egypt in the medium and long term, its immediate effect has been to bring Egypt near parity with Israel in terms of gross United States assistance per year.

But the rise in population of about 3 per cent per year in Egypt and the Sudan over a protracted period has meant that in order to stand still over the past twenty-five years it was necessary to expand the economies by *over* 3 per cent per year. While there were some years when the 3 per cent growth rate was substantially exceeded in Egypt, though not in the Sudan, the level of performance necessary to achieve really significant improvements in the standards of living of the mass of the population proved to be impossible over the period as a whole and some sections of society suffered falls in their standards of living. Such falls would have been greater if there had not in the case of Egypt been some fortunate external factors

Table 7.5   Economic indicators in Egypt and the Sudan 1986

Country	GDP($bn) GDP	per head	Growth (%) 1970–80	1980–8
Egypt	37.7	760	7.1	7.1
Sudan	7.3	240	2.2	0.2

Source: Un and IMF data quoted in *Encyclopaedia Britannica* (1990: 770–5).

contributing positively to the nation's economic performance (see Table 7.5).

By 1980 so important were these external factors, such as remittances from Egyptian professionals and workers living in oil-rich countries, the Suez Canal revenues, tourism revenues as well as the benefits of domestic oil production that the Egyptian economy experienced a brief period of balanced trade. Officially recorded remittances were estimated to total 10 per cent of national income by 1980 and unrecorded remittances were thought to be at least as much again. In other words remittances by Egyptians working overseas were as large as some of the major productive sectors, agriculture (20 per cent), industry (34 per cent). The important advantage of the official and unofficial remittance monies to both Egypt and the Sudan was that they were outstandingly favourable in net terms in that their value to the economy in purchasing power was immediate. Comparatively, the agricultural and industrial sectors generated as much or more turnover, but with inputs taken into account the immediate net gain to the economy was much less spectacular than that provided by remittances. The evaluation of remittances, however, should be more thorough than a mere comparison of the gross turnover of the major sectors of the economy in that there are many real costs to the Egyptian and the Sudanese economies associated with the short-term migration of nationals and especially with long-term migrants who return to Egypt to retire. Egyptian and Sudanese professionals and other workers have been educated, trained and have gained experience in their own countries. The economies which they served abroad avoided these economic and social costs. At the same time when they return these individuals and their families may be an economic and social burden on the Egyptian and the Sudanese societies and economies. Thus Egypt and the Sudan are servicing the heavy social costs of the youth and old age of their migrant workers and the respective national economies have very limited access to the benefits gained by their nationals working abroad. But there is no question of the Egyptian government intervening to impede the migration of its professionals and other workers in that the demographic pressures at home are so high that the ease provided by the migration process is perceived to be positive. The return of nearly 1 million Egyptians, including workers and their families, from Iraq and Kuwait as a consequence of the Gulf Crisis and the 1991 war confirmed the perception. The experience of 1980s and especially that of the 1991 Iraq–UN War demonstrated the vulnerability of the Egyptian and the Sudanese economies to regional political and economic disruption. The incomes of migrant workers are amongst the most vulnerable elements of the regional oil-based economy and as a result the nations providing skilled and unskilled labour suffer periodic shocks. The contraction of the economies of the Middle East oil producers from 1981 severely affected the volume of official, and the equally important unofficial, remittance monies during the 1980s and the trade gap widened and Egypt's debt position worsened. The position of the Sudan was relatively even worse. Remittances by Sudanese nationals had been important in the 1970s and they fell away painfully in the 1980s. Meanwhile the military disruption in the south of Sudan affected very negatively the economic performance of the country. The south itself declined in economic terms and the more populous north was affected by the diversion of resources to support the refugees caused by the war in the south, not to speak of the burden of military expenditure. At the same time the Sudan was obliged during the 1970s and especially during the 1980s to take on the burden of a foreign refugee population, exceeding 1 million people at times, fleeing from the conflict in Ethiopia. The situation was exacerbated by the drought of the early 1970s (Dalby and Church 1973) and the hardships of that period were then exceeded by the drought of the early 1980s and by that of 1990.

Both Egypt and the Sudan have been fortunate during the period of review in that world prices of major food staples such as wheat and livestock feed have been depressed by the competition of the two major food surplus producing regions, the United States and the European Community. Both the USA and EC have subsidized their agricultural sectors and have found it difficult to dispose of their food surpluses economically. Thus the Nile Basin countries have been protected from the worst effects of their inability to produce their food needs, although the possibility of cheap

Table 7.6 Volume and direction of trade in Egypt and the Sudan 1986

Country		Total ($bn)	Direction (%)				
			EC	US	USSR	Japan	Other
Egypt	Imports	11.5	38	15	13	5	29
	Exports	2.9	36	3	23	3	36
Sudan	Imports	1.4	38	9	2	3	48
	Exports	0.6	25	2	8	5	60

Source: Encyclopaedia Britannica (1990: 825–9).

imports does nothing to encourage the raising of internal producer prices.

The commodity trading performances of Egypt and the Sudan have not been strong. The Sudan had a 25 per cent commodity trade surplus in 1968, deriving from cotton (60 per cent of exports, gum arabic 10 per cent and groundnuts 7 per cent: *Encyclopaedia Britannica* 1990). At that time Egypt was already suffering a 10 per cent commodity trade deficit (see Table 7.6).

The majority of the trade, imports as well as exports, of both Egypt and Sudan is with the industrialized countries of North America, Europe and the Far East. These are the destinations of the major export commodities, cotton and other agricultural products and arms from Egypt, cotton and sugar from the Sudan. The industrialized world supplies the manufactured goods and equipment with which to develop and maintain the physical infrastructure of roads, railways, telecommunications, transport equipment and consumer durables. Only Egypt has established a significant manufacturing base, with the capacity to manufacture iron and steel and chemicals as well as fabricate and assemble some capital goods. Priority has unfortunately been given to arms manufacture, which has diverted energy from essential developmental investment. But it has to be said that the Egyptian economy was given an enormous boost by the Iran–Iraq War between 1980 and 1988. Iraq absorbed a large volume of Egyptian manufactured arms as well as up to 2 million Egyptian agricultural, industrial and service sector workers, who were needed to sustain the Iraqi war economy.

SOCIETY AND POLITICS

Both Egypt and the Sudan are mainly Muslim countries (see Table 7.7); both have been affected by the fundamentalist movements which suffered during the nationalist government of Nasser in Egypt but gained influence under the more indulgent administration of Sadat. In the 1980s Islamic fundamentalism increased in influence throughout the region and it caused the Sudanese President Nimeiri to reintroduce divisive Islamic legal conventions. The traditional and Christian population of the south, though a small minority in national Sudanese terms, reacted in such a violent way that the Sudan has been at war for over a decade, with terrible social and economic consequences. At no time in the 1980s was there a widely popular government in the Sudan.

Table 7.7 Religious beliefs in Egypt and the Sudan 1988

Country	Population 1988 (million)	Muslim (%)	Christian (%)	Traditional beliefs (%)	Other (%)
Egypt	50.3	94	6	—	0.1
Sudan	26.3	73	9	17	0.1

Source: *Encyclopaedia Britannica* (1990: 762–3).

After the experiment with a severe form of Islamic socialism in the 1960s under Nasser, Egypt has gradually liberalized its politics and economic institutions. The debate rages internally on the issues of centrally controlled versus free-market approaches to economic development. And Egypt's economic sponsors USAID, the UN and other international agencies such as the World Bank and the IMF, are liberal in their criticism of Egypt's low-priced food and fuel and widely subsidized economic system. Attempts to embark on a market strategy have been taken very slowly and there have been predictable and violent reactions especially during the 1970s under Sadat when food riots followed an increase in the price of bread. But for the most part a delicate balance has been struck between the need for reform and the political consequences of haste. Meanwhile the massive bureaucracy installed by Nasser and expanded by his successors has ensured that changes take place at a pace comfortable to Egyptians. It is possible to interpret the outcome as wildly inefficient in economic terms but if the intangibles of domestic accord and the effectiveness of the extraordinary redistributive mechanisms of the public sector are taken into account the result is something of an economic miracle of contained gradual expansion. Cairo itself is one of the most vigorous centres of urban expansion and construction in the Middle East, itself the major region of urban construction in the world during the period of the review. And Cairo has the most rapidly expanding urban construction sector in Africa.

The development of democratic institutions has been generally positive in both Egypt and the Sudan. Criticism of the government in Egypt is freely expressed although the tension between a liberal tending public administration and a fundamentalist Islamic opposition is palpable and the latter has been subject to suppression especially after the assassination of President Sadat in 1981 by fundamentalist sympathizers. The Sudan started well in democratic terms but the burden of the poor management of its indifferent natural resources has made the achievement of popular government very elusive. Despite the political circumstances and the war in the south elections were held in 1986 and more recently, but no individual or party has been able to command secure popular support.

REGIONAL AND INTERNATIONAL RELATIONS

External relations are important to both Egypt and the Sudan. Egypt has been so successful in cultivating the international community since the early 1970s that it has managed to survive grave deficit economic circumstances. The Sudan has been much less successful in attracting assistance and as a result its people have suffered. In terms of international assistance Egypt is in the big league, receiving up to $US3 billion annually in the late 1980s, in other words as much as Israel, although not receiving the level of aid per head enjoyed by Israel. The United States assistance to the Egyptian economy annually was running at the level of approximately 10 per cent of its GDP. In other words the continued growth of the Egyptian economy was dependent on a continuation of US support. No other country in the world has the motivation and the financial competence to sustain Egypt, a fact which cannot be ignored when considering the solidity of Egyptian support for the USA during the UN–Iraq War of 1991. Japan has the financial competence but not the motivation as Japan has no obligations to Arab countries of the sort which the United States has generated through its four decades of support for Israel.

It was very different twenty-five years ago. In 1965 Egypt and the United States had been through a decade of estrangement and Soviet support had been extended to assist in the finance of the building of the High Dam at Aswan. Soviet military equipment and advice was also extensive. But the flow of resources was not one way and the Soviet Union insisted that it be paid for its arms. The Egyptian economy had to produce the goods to pay for the Soviet assistance. By the early 1970s the need for substantial grain imports revealed that the Soviet Union could not supply Egypt's basic food needs and certainly could not provide the essential ongoing economic assistance. The departure of the Soviet advisers in 1973 came as no surprise to those who knew much about the Egyptian economy.

Sudan, meanwhile, which had attracted and sustained modest support from the United States, had during the period moved more towards policies which were designed to maintain the support of Islamic Arab sponsors from the Gulf and the Arabian Peninsula. After the increase in oil prices in 1973 considerable interest was shown in the Sudan by would-be Arab sponsors of agricultural and other development. These investments were boosted in 1979 and 1980 but the progressive collapse of the oil market and the poor performance oil Middle East and North Africa oil economies brought about a decline in Arab financial support during the 1980s and also very seriously reduced the level of remittance revenues.

The most significant change in regional politics followed the Camp David Accord of 1977, in which Egypt recognized Israel's right to existence and agreed to abide by non-military means to settle outstanding and future disputes. The relationship between Egypt and Israel did not as a result automatically become warm, and very few Egyptians have visited Israel in the years since 1977 although many Israeli tourists visit the rich antiquities of Egypt. The end to confrontation had a huge impact, however, on Egypt's relations with other countries. There were improvements in relations with the industrialized world and a parallel decline in those with Soviet bloc countries. Regionally the impact was devastating in that all Arab countries boycotted Egypt; the Arab League headquarters moved from Cairo to Tunis and diplomatic and other relations with the Arab World were frozen for ten years. By 1990 severed relations had been restored and Egypt rejoined the Arab community and was able to lend its support to its traditional concerns.

CONCLUSION

Egypt and the Sudan have a poor resource endowment and their economic prospects are further blighted by their levels of population increase. Egypt has been able to endure and even expand its economic performance as a result of a skilled use of its strategic position internationally. The Sudan has no such advantages and its economic, ethnic, religious and developmental difficulties have led to the virtual collapse of its institutions. The future for neither country is secure but the past performance of Egypt, both economically and particularly in the recent past by its political institutions, suggests that it will continue to expand, if slowly, while the worsening predicament of Sudan suggests that any improvements will come only in the very long term.

REFERENCES

Abate, Z. (1990) 'The integrated development of Nile Basin waters' in Howell, P.P. and Allan, J.A. (eds) *The Nile: Resource Evaluation, Resource Management, Hydropolitics and Legal Issues*, London: School of Oriental and African Studies, University of London-RGS.

Abdel Salam and Mohamed, M. (1976) 'Agriculture in the Sudan', in Al-Hassan, A.M., *An Introduction to the Sudan Economy*, Khartoum: Khartoum University Press.

Abdulla, I.H. (1971) 'The Nile Waters Agreement in the Sudanese–Egyptian relations', *Middle East Journal* 7(3): 329–41.

Al Ahram al Iqtisadi (1975) 'The High Dam and the campaign against it', *Al Ahram al Iqtisadi* (in Arabic) 1 February, Cairo.

Barnett, A. (1975) 'The Gezira Scheme: production of cotton and reproduction of underdevelopment', in Oxaal, I. and Booth, D. (eds) *Beyond the Sociology of Development*, London: Routledge & Kegan Paul.

Chesworth, P.M. (1990) The history of water use in Egypt and the Sudan, in Howell, P.P. and Allan, J.A. (eds) *The Nile: Resource Evaluation, Resource Management, Hydropolitics and Legal Issues*, London: School of Oriental and African Studies, University of London. pp. 41–52.

Collins, R.W. (1989) *The Waters of the Nile: Hydropolitics and the Jonglei Canal, 1900–1988*, Oxford: Oxford University Press.

Collins, R.W. (1990) 'History, hydropolitics and the Nile: Nile control, myth or reality', in Howell, P.P. and Allan, J.A. (eds) *The Nile: Resource Evaluation, Resource Management, Hydropolitics and Legal Issues*, London: School of Oriental and African Studies, University of London-RGS.

Dalby, D. and Church, H. (1973) *Drought in Africa*, London: School of Oriental and African Studies, Centre of African Studies.

Dykstra, D.I. (1977) 'A biographical study in Egyptian modernization: Ali Mubarak – 1823/4–1893 – Volume 1' PhD dissertation, University of Michigan, Ann Arbor: University Microfilms.

Encyclopaedia Britannica (1990) *Book of the Year*, Chicago.

Evans, T.E. (1990) History of Nile flows in Howell, P.P. and Allan, J.A. (eds) *The Nile: Resource Evaluation, Resource Management, Hydropolitics and Legal Issues*, London: School of Oriental and African Studies, University of London.

FAO (1973) *Perspective Study of Agricultural Development for the DRS: Land and Water Development and Use*, ESP/AGL/PS/SUD/73/6, Rome: FAO.

Gaitskell, A. (1959) *Gezira: A Study of Development in the Sudan*, London: Faber.

Hopper, W.D. (1976) 'The development of agriculture in developing countries', *Scientific American* 235(3): 197–205.

Howell, J. (1973) 'Politics in the Southern Sudan', *African Affairs* 72(287): 163–78.

Hulme, M. (1990) 'Global climate change and the Nile Basin', in Howell, P.P. and Allan, J.A. (eds) *The Nile: Resource Evaluation, Resource Management, Hydropolitics and Legal Issues*, London: School of Oriental and African Studies, Centre of Near and Middle Eastern Studies.

Hurst, H.E., Black, R.P. and Simaika, Y.M. (1965) *Long-Term Storage: An Experimental Study*, London: Constable.

Little, T. (1965) *High Dam at Aswan*, London: Methuen.

Ministry of Agriculture (1990) Personal communication – London Nile conference, London: School of Oriental and African Studies, University of London-RGS.

O'Brien, P.K. (1966) *The Revolution in Egypt's Economic System*, London: Oxford University Press.

Okidi, O. (1990) 'A review of treaties on consumptive utilization of waters of Lake Victoria and Nile drainage basins', in Howell, P.P. and Allan, J.A. (eds) *The Nile: Resource Evaluation, Resource Management, Hydropolitics and Legal Issues*, London: School of Oriental and African Studies, University of London-RGS.

Permanent Joint Technical Commission (1970–87) Technical data, Cairo and Khartoum: PJTC.

Republic of the Sudan, Ministry of Irrigation (1955) *The Nile Waters Question: The Case for Egypt and the Sudan's Reply*, Khartoum.

Shahin, M.M.A. (1985) 'Hydrology of the Nile Basin', *Developments in Water Science, 21*, Amsterdam: Elsevier.

Stoner, R.F. (1990) 'Future irrigation planning in Egypt', in Howell, P.P. and Allan, J.A. (eds) *The Nile: Resource Evaluation, Resource Management, Hydropolitics and Legal Issues*, London: School of Oriental and African Studies, University of London-RGS.

Sutcliffe, J.V. and Parkes, Y.P. (1987) 'Hydrological modelling of the Sudd and Jonglei Canal', *Hydrological Sciences Journal* 32: 143–59.

Waterbury, J. (1979) *The Hydropolitics of the Nile*, Syracuse, NY: Syracuse University Press.

Wickens, G.E. (1982) 'Paleobotanical speculations and Qaternary environments in the Sudan', in Williams, M.A.J. and Adamson, D.A.A. (eds) *Land between the Two Niles*, Amsterdam: Balkema.

Whittington, D. and Haynes, K.E. (1985) 'Nile water for whom? Emerging conflicts in water allocation for agricultural expansion in Egypt and Sudan', in Beaumont, P. and McLachlan, K.S. (eds) *Agricultural Development in the Middle East*, London: John Wiley.

Wynn, R.F. (1971) 'The Sudan's Ten-Year Plan of economic development, 1961–62 – 1970–71. An analysis of achievement to 1967–68', *Journal of Developing Areas* 5: 555–76.

8

THE TROUBLED ARAB MIDDLE EAST

St John B. Gould

INTRODUCTION

The heart of the Middle East is the focus of this chapter. It includes the Arabian Peninsula and Fertile Crescent regions of the Mashreq (the eastern part of the Arab world), the states of Yemen (formerly Yemen AR and Yemen PDR), Oman, the United Arab Emirates (UAE), Qatar, Bahrain, Saudi Arabia, Kuwait, Iraq, Syria, Lebanon and Jordan (see Figure 8.1). This area is characterized by the predominance of Islam, the identification of Arabic-speaking people with an Arab cultural and ethnic identity, the pervasive economic influence of oil and natural gas production, arid environments and widespread political turbulence and conflict which has constantly involved outside powers.

POPULATION: GROWTH AND MIGRATION

In general the Arab Middle East is sparsely populated, with all except Lebanon and the small Gulf states of Bahrain, Kuwait and Qatar having population densities of less than 65 persons per square kilometre. This can be explained in part by the region's arid environments. However, population densities are higher in areas of intensive farming, such as in central Iraq, and in the urban agglomerations. Thus the northern states of Iraq (in settled times), Jordan, Lebanon and Syria have a population density of 46 persons per square kilometre, as opposed to 7.5 in the Arabian Peninsula (Blake et al. 1987).

Over the last twenty-five years population growth has been particularly rapid (Table 8.1). Saudi Arabia's population has

191

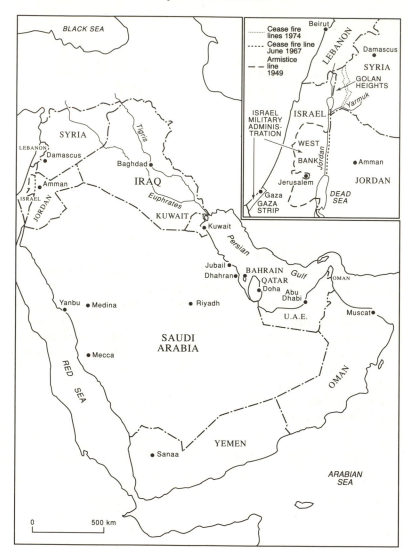

Figure 8.1   The Arab Middle East and Israel

increased from under 9 million in 1979 to over 14 million in 1988 and is also projected to double between 1990 and 2010. Similar patterns are evident in other countries. Annual population growth rates are outstanding in global terms. In the period 1984–9 Qatar's rate of 6.4 per cent is the highest of the world's sovereign states, the

Table 8.1   Population growth in the Arab Middle East 1979–88

	Population 1979	Population 1988	Births per 1,000	Deaths per 1,000	Natural increase per 1,000
Bahrain	331,000	481,000	36.8	5.8	31.0
Iraq	12,821,000	17,657,000	45.1	8.6	36.5
Jordan	2,842,000	3,943,000	34.7	5.8	27.9
Kuwait	1,290,000	1,958,000	26.2	2.2	24.0
Lebanon	2,696,000	2,828,000	28.0	7.0	21.0
Oman	935,000	1,378,000	44.2	13.0	31.2
Qatar	217,000	341,000	30.4	2.4	28.0
Saudi Arabia	8,931,000	14,016,000	37.3	12.8	24.5
Syria	8,421,000	11,338,000	44.1	7.0	37.1
UAE	919,000	1,501,000	33.5	3.9	29.6
Yemen AR	5,837,000	7,535,000	49.1	20.8	28.3
Yemen PDR	1,910,000	2,514,000	48.0	15.0	33.0

Sources: Totals from United Nations (1988); rest from *Encyclopaedia Britannica* (1990).

UAE's 5.1 per cent the second highest, whilst Kuwait and Saudi Arabia are in fourth and sixth place respectively. Yet it should be noted that all four states have imported labour from other states, both within and outside of the region, and that the natural rate of increase is less than that of Kenya. The only state in the region to have experienced population growth of under 2 per cent, Lebanon, suffered considerable emigration during the 1980s due to its protracted civil war and the intervention of Syrian and Israeli armed forces. It is too early to say exactly what has been the impact of the Iraq–UN War: the returning government of liberated Kuwait has indicated that it wishes permanently to reduce the population (in the past it was in the majority non-Kuwaiti), and there is no doubt that a massive exodus took place. Similarly, there have been massive exoduses from Iraq – but the long-term nature of these movements is not yet clear. Historically, the 1967 Arab–Israeli War and the occupation of the Gaza Strip and the West Bank by Israel is another example of armed conflict which added great numbers of refugees to those already living outside Palestine.

Birth rates are high, as in much of the developing world, and birth control measures are poorly received. At the same time death rates are equivalent to those of a number of European states. Thus all except Saudi Arabia, Oman and the two parts of Yemen have a lower death rate than the United Kingdom, due mostly to advances in health care financed by oil revenues. Infant mortality rates have fallen in most countries, although in some, notably Yemen, they are still high.

Large-scale international migration of labour has of course been prompted by disparities in per capita incomes between the various states (Al-Moosa and McLachlan 1985). The oil-exporting states of the Gulf required a substantial immigrant labour force following the 1973–4 oil price rise, and the subsequent development boom created employment opportunities that the small domestic labour forces could not satisfy in numerical or skill terms. Thus the foreign labour force of these states more than doubled between 1971 and 1975, and reached 4.3 million (28 per cent of the labour force) in the Arab world as a whole, by 1985.

Labour-supplying states in the region have begun to recognize the disadvantages the process brings. Although remittances are a source of national income and unemployment is reduced, migration is selective in that it depletes the higher skill levels, and detracts from the most productive age groups. Thus it has had a negative effect on agricultural production in Yemen AR, for example, and Jordan, which had 40 per cent of its labour force working overseas by 1975, and had to import 130,000 replacement workers by 1982. Remittances and the changing expenditure patterns of migrant workers have had an inflationary effect on the economies of labour suppliers such as Jordan (Birks and Sinclair 1980; Blake et al. 1987).

URBANIZATION

The Arab Middle East is a region where urbanism has historically been of great significance, with Mesopotamia (Iraq) and other areas giving rise to some of the world's earliest cities. The last twenty-five years, however, have witnessed a rapid process of urbanization, as modern economic development, stimulated by increasing oil revenues, is concentrated in urban areas. The period 1970–5, for example, reveals particularly high urban population growth rates, with Syria at 4.2 per cent, Lebanon at 4.9 per cent, Iraq at 5.0 per cent, Lebanon and Yemen PDR at 5.4 per cent, Saudi Arabia at 6.3 per cent, Yemen AR at 8.0 per cent and Kuwait at 8.2 per cent (Clarke 1980). The result is highly urbanized societies (Table 8.2). By way of comparison, Bahrain, Qatar and the United Arab Emirates are now more urbanized than either the United States or France.

Table 8.2 Urban population and GDP in the Arab Middle East

	Date	% Urban	GDP (US$ mn)	per capita GDP (US$)	GDP growth (%)
Bahrain	1981	80.7	2,670	8,530	2.8
Iraq	1987	70.2	34,000	1,950	1.7
Jordan	1979	59.5	4,370	1,540	3.5
Kuwait	1985	100.0	27,324	14,870	
Lebanon	1970	60.1		1,492	
Oman	1985	8.8	7,768	5,780	1.6
Qatar	1985	88.0	4,129	12,360	0.4
Saudi Arabia	1985	65.9	74,000	5,480	0.0
Syria	1981	47.0	20,421	1,820	9.3
UAE	1981	80.8	22,827	15,680	2.1
Yemen AR	1981	10.2	4,918	580	2.4
Yemen PDR	1973	33.3	956	420	1.2

Sources: Urbanization: *Encyclopaedia Britannica* (1990); economic data: (refers to 1986 or 1987) *Middle East Review* (1990).

Rapid urban growth in the Arab Middle East presents a wide variety of challenges for planners and administrators. First, as with most of the developing world the rapid urbanization process has led to deficiencies in housing and employment, and also in vital services and infrastructure in urban areas. This affects the poorest sections of the population, and the most recent migrants, most dramatically. The housing crisis has had two consequences: the concentration at great densities of a large and poor population in the urban core, and the evolution of shanty settlements on the peripheries of many of the region's cities where residents have poor access to services and facilities (Clark 1980). In some cities, for example Amman in Jordan, this latter problem is compounded by the refugee camps that house Palestinians who have left or been deported from Israel. Solutions to the problem of shanty settlements have included their removal, which does not solve the housing shortage and the provision of low-cost housing, the rent for which is usually beyond the means of the shanty dwellers. One result of urban growth has been increasing social segregation along economic class lines and the unplanned spread of middle-class suburbs emphasizes the need for infrastructural development in cities in the region, such as Beirut.

Many of the Arab Middle East's cities contain areas of great antiquity, where the buildings and water, drainage and sewage systems cannot cope with rising population densities and modern economic activities, and where the road network is insufficient for higher levels of traffic (Lawless 1980; Lewcock 1986). The need for the expansion of existing towns and cities and the building of new urban areas focuses the debate on the validity and relevance of starkly different Islamic and European models of urbanism and architecture. Great variations in social, urban and physical environments create quite particular challenges for the planners and architects, challenges which have been met with varying degrees of success. In Saudi Arabia the dichotomy is particularly distinct. There the oil industry has necessitated the creation of new industrial centres on a Western model, such as Yanbu and Jubail, and oil towns to house the employees of the international oil companies, such as Dhahran, all names which featured prominantly in the Iraq–UN War. Meanwhile, the role of the state as the custodian of Islam's most important shrines has created an altogether different set of circumstances for planners. This is most notable with the development of Mecca to cater for the large number of pilgrims, and the extension to the city's main mosque (Kemp 1990).

ECONOMIC CHANGE

Economic change in the Arab Middle East has been most noticeable in the oil-exporting states of the Gulf (Table 8.3) and it is the economies of these states which have had the profoundest effect on the region as a whole. There is a clear division between those states dominated by revenues from oil exports, which have high per capita incomes, low agricultural potential and productivity, small populations and an extensive reliance on immigrant labour (Kuwait,

Table 8.3   Oil in the economies of the Arab Middle East

	1989 Proven crude oil reserves ('000 million barrels	Value of petroleum exports (US$ millions)		
		1965	1979	1989
Iraq	100.0	660	21,382	14,500
Kuwait	94.5	1,207	16,770	10,863
Oman	4.3	—[a]	—[a]	—[a]
Qatar	4.5	141	3,643	2,000
Saudi Arabia	255.0	1,073	62,855	24,000
UAE	98.1	190	12,915	11,500

Source: OPEC (1990).
Note: [a] Data unavailable.

Table 8.4 Agriculture and industry in the economies of the
Arab Middle East

	Labour in agriculture, fishing and forestry as % of active work-force[a]	Industry % of GDP[b] 1973	1985
Bahrain	2.0	15.0	11.4
Iraq	27.7	9.9	11.6
Jordan	6.2	—[c]	—[c]
Kuwait	1.9	4.9	8.0
Lebanon	19.1	—[c]	—[c]
Oman	23.3	0.4	3.3
Qatar	0.1	2.0	8.2
Saudi Arabia	14.3	6.0	8.3
Syria	30.0	—[c]	—[c]
UAE	5.0	2.0	9.8
Yemen AR	59.9	—[c]	—[c]
Yemen PDR	46.6	—[c]	—[c]

Sources: [a] *Encyclopaedia Britannica* (1990) [b] McLachlan and McLachlan (1989).
Note: [c] Data unavailable.

Table 8.5 Farm area and food imports in the Arab Middle East

	Farmland as % land area Date	%	Cereal Imports ('000s metric tonnes 1974	1988
Bahrain	1980	5.2		
Iraq	1971	13.1	870	4,442
Jordan	1983	4.1	171	874
Kuwait	1985–6	0.3	101	417
Lebanon	1980	27.0	354	537
Oman	1978–9	0.3	52	293
Qatar	1986	5.7		
Saudia Arabia	1983	1.0	482	5,179
Syria	1987	33.1	339	1,044
UAE	1978	0.2	132	458
Yemen AR	1977–83	0.1	158	754
Yemen PDR	1977	0.1	148	459

Bahrain, Qatar, the United Arab Emirates, Oman and Saudi
Arabia), and those states with non-oil-dependent economies, where
agriculture is more important (Tables 8.4 and 8.5), labour more
available and per capita incomes lower (Syria, Jordan, Lebanon and

Yemen). Iraq is the exception to this pattern, as agriculture is important, the labour force is relatively large, but the state gained considerable revenue as a major oil-exporting state before the imposition of the UN trade embargo in August 1990, which one presumes is a temporary measure.

The oil economies

The transformation of the economic geography of the Arab Middle East in the period 1965–90 owes much to the changes in and increased importance of oil and natural gas (hydrocarbons) production (see Figure 8.2). The oil-producing states of the Arab Middle East, principally Iraq and the monarchies of the Gulf Co-operation Council, possess a majority (56 per cent) of the world's reserves of oil (OPEC 1990), which explains, at least in part, the international community's perception of the area as strategically important, and the increased power of these states themselves. Beyond the geopolitical sphere, the rise of the oil industry has had considerable impact in the countries concerned, releasing funds for investment in industrialization programmes, agricultural development, infrastructure, services and the welfare state, and has led to a dramatic increase in the average standard of living (McLachlan and McLachlan 1989). Most importantly it has created a situation of economic dependency on oil revenues rather than, as intended, a process of sustainable economic development. The impacts extend even beyond the oil-producing countries themselves, affecting the remaining states of the region.

In spite of sustained production levels the known oil reserves of the Middle East have increased significantly. Thus in 1965 the Middle East, including Iran, held 214.8 million barrels out of the world's reserves of 353.2 million barrels, and in 1989, 660.2 million out of 1,005.6 million, whilst in the period 1969–89, 66.3 per cent of the increase in the world's proven oil reserves was recorded in Middle Eastern countries (OPEC 1990; BP 1990). Whilst the Middle East's strong oil endowment indicates the considerable impact that the oil industry will continue to have on the region's economic geography, it is the evolution in production patterns that has had particular influence over the last twenty-five years. The importance of the region in global terms was similar in 1989 to that of 1965, 12,707,200 barrels per day out of 58,637,600 (21.7 per cent), as opposed to 6,454,000 barrels per day out of 30,201,500 (21.4 per

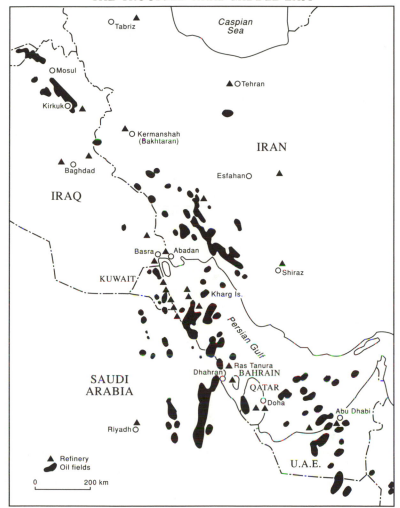

Figure 8.2   Location of oil fields in the Arab Middle East

cent). However, this proportion was higher (29.2 per cent) at the peak of the region's production in 1979, with levels of 18,360,500 barrels per day. More significant, however, is the changing balance of production within the region.

One state, Bahrain, has almost ceased to have any significance at all, continued export production relying largely on the generosity of Saudi Arabia in the shared exploitation of an offshore field. The production of Qatar and Oman is of a low level, as is the more

199

recently developed oil industry of Yemen. Over the last twenty-five years Saudi Arabia has become the dominant producer in the Middle East, acting as the pivotal influence on prices within OPEC, and assuming a strategic significance in the international economy. Within the Arab Middle East, the other oil power is Iraq. However, its production was severely affected by the damage to its facilities and transportation networks caused by the war with Iran from 1980, but 1988 and 1989 saw it regain levels of over 2,700,000 barrels per day, before the massive curtailment caused first by the UN embargo of 1990–1 and then by damage to refineries during the following war over Kuwait. At the beginning of the period Kuwait was the region's largest producer, but the conservative oil policies of its government in the 1980s reduced its importance, although not enough to appease Iraq and to raise world prices, even before the horrific sabotage by Iraq during the 1991 war curtailed production completely in the immediate term. The most dramatic rise has been in the United Arab Emirates where production has increased by over 658 per cent between 1965 and 1989 (Table 8.3).

The Organization of Petroleum Exporting Countries (OPEC), the focus of which is now the Gulf states of Iran, Iraq, Kuwait, Saudi Arabia and the United Arab Emirates, had been established by the first four and Venezuela in 1960 to co-ordinate relations with the international oil industry, in order to gain higher and more stable prices. Control of the oil industry was most decisively wrested from the oil companies when the Arab–Israeli War of 1973 resulted in an Arab oil embargo on the USA and the Netherlands, and the quadrupling of prices. In the years following the price shock the oil-producing countries of the region enjoyed a boom period when revenues enabled considerable investment, to the point of exceeding the absorptive capacity of the economies of some states (McLachlan 1980). During the 1980s oil revenues have slipped in real terms as the OPEC countries have failed to maintain their share of world production and thus their control over prices, whilst the problem of economic dependency on oil, created in the mid- and late 1970s, has only increased as the revenues have diminished.

It has often proved the case that the availability of a relatively easy income from oil has diminished the capacity of both government and people for serious competitive labour, and that the viability of employing foreign labour, indeed the necessity of using it in a period of rapid development when all the oil-rich states in the region except Iraq have chronically insufficient human resources, prevents the domestic labour force from developing a wide range of work skills. Thus in 1980 a majority of the work-force of Kuwait, Qatar, Saudi Arabia and the United Arab Emirates were immigrants. The demand for labour in the post-war reconstruction of Kuwait will be very high: whether or not it will also be high again in Iraq depends on the extent to which that country demobilizes man-power from its swollen armed forces in the wake of military humiliation.

The most successful programme of investment by an individual state in the region, Kuwait, has not been in the industrial or agricultural sectors, but in overseas financial institutions, the proceeds from which now earn the state more than oil production does, and the future of which is more sustainable: it is of course notable that these were assets that lay beyond the grasp of the invading Iraqis.

The difficulty with assessing the extent of oil dependency within individual economies is in ascertaining the proportion of the non-oil economic activity which is dependent on either oil revenues or on cheap oil supplies for its continuation. Taking Kuwait as an example, it is estimated that between 78 per cent and 86 per cent of the pre-war national income resulted from the impact of oil on the economy directly of overseas revenue investment. Oil formed a similar high proportion of export earnings for other states, in 1986 92 per cent for Oman, 94 per cent for the United Arab Emirates, 95 per cent for Iraq and 99 per cent for Saudi Arabia. Collapses in oil prices have therefore resulted in large-scale borrowing, as expenditure has not been adjusted to take into account the fall in available foreign exchange. As a consequence, of the oil-exporting Gulf states, only Bahrain, due to the decline in its oil industry and the development of trade and banking, and Iraq, because of its larger population and greater agricultural potential, can claim in any way to have mixed economies. Table 8.4 shows the small place of industry in these economies, even although activities such as oil refining are included.

Saudi Arabia's industrial sector contributed only 4.6 per cent of its GDP in 1985, and refining and petrochemicals constituted 96 per cent of the value of its manufacturing sector. Saudi Basic Industries Corporation (Sabic), which produces petrochemicals, plastics, steel and fertilizers, earned about US$800 million profits in 1988 (Kemp 1989a). Indeed, the process of diversification away from oil has tended to be problematic for the region as a whole.

ST JOHN B. GOULD

The non-oil economies

Those states of the Arab Middle East which are not major oil-exporters are characterized by many of the problems that beset other regions of the developing world, but their proximity to the oil-states has produced specific features as well. Jordan and Yemen AR have become major exporters of labour to the oil economies, whilst oil refining dominates the industrial economy of Yemen PDR. Since 1984 both Syria and Yemen AR have had oil discoveries, so that in both Syria and unified Yemen today oil is the most dynamic economic sector. It may be argued that these too are becoming oil dependent, though post-dating the oil boom. In all of the states manufacturing forms a relatively low proportion of GDP.

The Lebanese national economy is almost non-existent, due to the civil war, and although a surprising level of economic activity persists, the state has run a trade deficit during the 1980s. Syria's GDP fell 9.3 per cent in 1987 and the per capita rate has fallen consistently since 1982, whilst inflation in 1989 was above 100 per cent. Industry is relatively well developed, expanding by 23 per cent between 1985 and 1987, but defence spending remains the principal burden on an economy traditionally propped up by Arab and (now negligible) Soviet aid. Jordan's problems in continuing economic development are centred around its debt, with recent IMF imposed austerity measures leading to great social unrest. Phosphates are a dominant element in the economy, but attempts have been made to spread industrial development. The short-term effects of the Iraq–UN War are well-known: its earnings from acting as a entrepôt for Iraq have collapsed, while it has also had to cope with a collapse in remittances and a flood of refugees. Yemen's economy is in transition, the south recovering from the political crisis of 1986 and the north adjusting to large-scale oil production, and the whole in a process of integration which is in its formative stages.

AGRICULTURAL CHANGE

For most states of the Arab Middle East agriculture has ceased to be a central economic sector. Its share of GDP has in general fallen, and although labour participation is still relatively high in some states (Table 8.4), and although rates of agricultural production have grown, dependency on cereal imports has increased. Agriculture remains important as an employer in all states except Jordan, Bahrain, Kuwait, Qatar and the United Arab Emirates, and is most significant in Yemen.

The problem of reliance on imported cereals is most critical in Iraq and Saudi Arabia, the latter recently investing heavily in agricultural development. In all states this dependency has worsened, as populations have expanded. Whilst the food deficit is affordable at present for the oil-exporting states, for they are the highest spenders on food imports (Weinbaum 1982), if and when oil revenues decline a more serious problem will have to be faced. In only one country, Yemen, do nutritional levels mark it out as a 'food priority country', as identified by the World Food Council. Yet should the food deficit grow, particularly in poorer states such as Syria and Jordan, low nutritional levels could become more common. The Middle East and North Africa is at present the region of the world in which the food deficit is growing most rapidly. Whilst cereal yields have made dramatic gains, their relative levels in world terms have decreased. Moreover, the problem is not just one of population growth, but of rising per capita demand. A United Nations estimate indicates that production in the Middle East and North Africa will have grown by 148 per cent from 1970 to the year 2000, but demand by 227 per cent. In this region potential self-sufficiency from irrigated or dryland agriculture is estimated to include at the most Syria, Lebanon, Iraq and Yemen AR. There are two elements in the struggle to increase production; the natural constraints imposed by the environment, and improvements made in spite of the environment through the development of technology and techniques, and through massive investment. The most important constraint on agricultural production in the Arab Middle East is the availability of water (Beaumont et al. 1988; Allan 1985). In all states other than Syria, Lebanon and Iraq, farmland forms a very minor part of total land use (Table 8.5).

Rain-fed agriculture requires precipitation of at least 240 mm of rainfall per annum, and a low inter-annual variability, yet to be commercially successful the required figure is much higher. With the characteristic high temperatures of the region, evapotranspiration rates are also of critical importance. Thus in this region such activity is limited to small parts of south-western Saudi Arabia and northern Oman, to Yemen AR, Lebanon and parts of Iraq, Syria and Jordan. The extension of agriculture beyond the areas of relatively high rainfall is dependent on irrigation, which has historically been of significance in the Middle East, with the existence of well-established and complex traditional systems such as the *Qanat* (traditional systems of small underground canals which tap the local water-table). Rates of fertilizer and pesticide application have increased dramatically, but without the extension of croplands can do little to increase production significantly. For the poorer countries of the region the development of agriculture has been a slow process. However, the extent to which rapid development can be attained if sufficient funds are applied is illustrated most obviously by Saudi Arabia, the agricultural output of which tripled in value between 1984 and 1989 to more than US$6,000 million, which exports wheat, and which is almost self-sufficient in dairy produce and vegetables (Kemp 1989b). Thus wheat production increased from 300,000 tonnes in 1976 to 2.8 million tonnes in 1988, 1.6 million tonnes of which was sold outside the Gulf Co-operation Council. In 1987 wheat was exported at 30 per cent of its cost price, and thus a long-term export strategy is unlikely to be economical. Central pivot irrigation has extended cropland to previously arid areas, whilst farmers have been supported by an extensive backup of free land, interest-free loans, subsidized fertilizers, feed and equipment, and guaranteed prices for wheat. These radical agricultural developments have helped to diversify the economy away from oil and slow down the process of rural–urban migration, but have required substantial and uneconomic investments and the use of 89 per cent of the kingdom's water consumption annually. In consequence aquifers are being depleted and becoming steadily more expensive to exploit.

TRADE AND AID

Trade patterns in the Middle East are heavily orientated towards the West, the market for most of the region's oil exports and the source of much of its manufactured goods and food imports. The region suffers in relation to the West because of its reliance on the export of primary products, making it vulnerable to price instability. This is the case not only with the crude oil-exporting states of the Gulf, but also with Jordan, which relies to a large extent on the export of phosphates, the price for which often fluctuates wildly, as in 1976–7. The reliance of the region on the West exposed it to the preference of that market for crude oil rather than refined oil or oil products, yet as they are an important export market for the European Community the Arab states do have some bargaining strength. Those states with agricultural commodities to export, notably Syria, Jordan and Lebanon, suffer from the protectionism of the European Community, a situation which is likely to increase after the development of the single market in 1992. More recently, trade has increased substantially with the Far East, due to the competitiveness of its manufactured goods and the need of states such as Japan and South Korea for oil. There has been little success in efforts to improve trade within the region. The Arab Common Market, established by the Arab League in 1965, has had little impact, due to the rejection of full membership by all states other than Iraq, Syria, Jordan and Egypt, and because the members had little they could trade with each other. Aid flows from the oil-rich states have been relatively large as a proportion of GDP, but small in relation to overall funds available. There was considerable disbursement of aid during the 1970s oil boom, to Arab, Islamic and Third World states, but in the 1980s as prices fell, so did disbursements. Those from Saudi Arabia, Kuwait and the United Arab Emirates fell from US$7,940 million to US$1,365 million between 1980 and 1989, and ironically the Iraqi war effort took up much of the assistance, since Saudi Arabia and Kuwait in particular, both conservative monarchical Arab countries, feared the potential export of Islamic revolution from Persian Iran. The weaponry was turned savagely against the paymasters in 1990–1, and they in turn deflected this kind of support to add to donations from the EC, Japan and USA to the front-line states most affected by the Iraq–Kuwait crisis, principally Jordan (which later compromised itself in Saudi and Kuwaiti eyes), Egypt and Turkey (Middle East Economic Digest 1990). In general, Arab official development assistance, which declined anyway in the 1980s, may now be realigned in very different ways – although the fall-out from the war is still not clear. Most of that aid went to Egypt, Pakistan and Sudan, the only major recipients in the Arab Middle East being Yemen, Jordan and Syria.

ENVIRONMENTAL ISSUES

As has been indicated elsewhere, agricultural development has been extensive in parts of the Arab Middle East in recent years. It has produced locally significant changes in land use, notably with the extension of irrigation in valley and plain lands, whilst the abandon-ment of uplands threatens landform collapse, notably in Yemen. Furthermore the degradation of land, such as through erosion and salination is an extensive process in the Middle East. An example of the latter problem is in Iraq, where 60 per cent of irrigated land suffers from some sort of salinity, seriously reducing agricultural productivity.

As with many areas of the developing world pollution is becom-ing a critical problem in the Arab Middle East (Morad 1990). Green politics are stifled in many of the states, but awareness was growing even before the horrors of the world's biggest ever oil slick in the Gulf and the torching of Kuwait's oil fields, both acts of unparalleled sabotage. The problem for many states is that restric-tive environmental planning and pollution controls are seen to conflict with the objectives of economic growth. Thus Saudi Arabia has rules for the monitoring and punishing of infringements that other states cannot afford (Parker 1990). In the Arab Middle East air pollution before the Iraq–UN War consisted principally of primary emissions from the burning of fossil fuels, from car exhausts, factory chimneys and household use, and from the interaction of two or more primary contaminants and atmospheric constituents. The climate is particularly susceptible to smog formation, and there has been a consequent rise in related medical complaints. Smog has the potential to become a catastrophic problem even in normal times, threatening Baghdad, Amman and Kuwait City in particular. The abundance of fossil fuels in the region makes energy conserva-tion programmes difficult to enforce, whilst new non-traditional building materials are generally less suited to the environment and lead to greater energy consumption.

Water pollution is also on the increase, in rivers and ground-water aquifers, and the problems will be exacerbated as population growth and industrial demand use up remaining uncontaminated water resources. Water pollution tends to involve excreta, water-borne diseases, organic industrial waste, fertilizers or pesticides. Despite the rehabilitation of some river stretches through control-ling measures, as in the Tigris–Euphrates system, aquatic life is often endangered (Naff and Matson 1984). The grave crisis over water resources has led to a number of trends: the development of research into aquifers and arid environments, and the construction of desalination plants, notably in Kuwait, Qatar and Saudi Arabia, some of which hit the headlines in early 1991 when the massive oil pollution in the Gulf threatened the saltwater intakes. The problem with this latter strategy is that it is still very energy expensive.

In the Gulf in 'normal' times oil slicks can be a problem. Given the shallow nature of the sea and its restricted connection with the Arabian Sea and Indian Ocean, pollutants are not dispersed quickly, and can have a rapid impact on the fishing grounds. Success in normal times in limiting pollution has been slow and difficult to achieve, as in Iraq, Kuwait, Qatar, Bahrain and the United Arab Emirates, notably with the banning of the burning of timber and coal. The extraordinary damage caused by sabotage during the Iraq–UN War has highlighted the problems of the area.

POLITICAL CHANGE

This has three themes: the struggle for unity, as in the United Arab Emirates, Yemen, and regional groupings such as the League of Arab States and the Gulf Co-operation Council: conflict, principally focused on Lebanon, Israel/Palestine and the Gulf; and the evolution of boundaries, which in part explains and reflects some of the conflicts.

Unification is represented by the formation of the United Arab Emirates in 1971, with Ras al-Khaimah completing the present federation in 1972 (Rumaihi 1986). The original six emirates were all British Protectorates, but that on its own was not a guarantee that unification would follow. The UAE and the two former British Protectorates of Qatar and Bahrain have since developed as relatively stable and wealthy Western-orientated states. Oman appears in a similar light, although only after surviving a revolutionary struggle in 1970–5. Yemen PDR (formerly the British Protectorate of Aden), which gained independence in 1967, was dominated by a succession of leaders with strong socialist policies who developed close ties with the USSR. After the ending of its civil war in 1970, the other (North) Yemen, previously a feudal monarchy, was ruled by republicans with the increasing involvement of the military. In November 1989 the two Yemens announced plans for unification, and in 1990 they merged to form a single republic (Edge 1990), with its capital in Sana'a (Yemen AR) and unified political, administrative and military institutions. In many senses Yemen AR is dominant in this arrangement with a larger population, greater oil reserves, a more stable economy, the retention of most of its political and economic character and the appointment of its President as the first of the united republic. Yemen PDR, with economic problems and the recent loss of substantial Soviet aid, is the lesser component of the two. However, the unification, following that of the United Arab Emirates in 1971 and the establishment of the Gulf Co-operation Council in 1981, is a sign that the process of unification is not universally in retreat in the Arab Middle East.

The possibility that countries which are riven internally by religion, sect and tribe will descend into anarchy and civil war, as in Lebanon, generates the kinds of circumstances in which dictator-ship can also flourish, as in Syria and Iraq. Such dictatorships may claim legitimacy in terms of a unifying ideology such as a brand of Arab socialism, but rarely do they weld into a co-operative inter-national community. Rather, they bring national aspiration and rivalry to the fore. For this reason states can appear to espouse the same creed, but be implacably oppposed to each other. Ba'athism, a poorly defined blend of Arab nationalism and socialism, has dominated Syria from 1963 and Iraq from 1968. Syria has been ruled in an autocratic fashion by Hafez Assad from 1970, who has presided over a major expansion of the country's armed forces, with Soviet aid, which were used in conflict with Israel in 1967 and 1973, and in Lebanon since 1976. The challenge from fundamentalist Islamic groups has posed a threat, notably leading to the massacre of several thousand people in Hama in 1982. Iraq under Saddam Hussein, who has ruled since 1980, expanded the power and size of its armed forces, so that in the run-up to the Iraq–UN War of 1991 it was frequently described as having the world's fourth largest army – and that in a country with only 17 million inhabitants. The army fought a long and ultimately futile war with Iran during the 1980s, principally over border issues, but also building on centuries-old emnity between Arab and Persian. The fragility of Arab unity was exposed by the fact that Syria (and Egypt) committed armed forces alongside those of the USA, Saudi Arabia and Kuwait and other UN states to the struggle to evict Iraq from Kuwait. Syria emerged with a much stronger regional position as a result of Iraq's defeat.

The longest running conflict has of course been over Israel, whose right to exist has been accepted in the region only by Egypt. The recognition of the state has been held in abeyance principally because during its original foundation and several times since, Arabs (in what was originally known as Palestine), have been evicted from their lands and homes. These refugees form the seed-bed from which the Palestine Liberation Organization draws its support and they are to be found scattered over many countries of the Arab Middle East. They formed a large part of the immigrant population of Kuwait pre-1991. Jordan lost the West Bank to Israeli military occupation in 1967, creating further refugee problems, and Palestinians have since played an important although controlled role in the political life of the East Bank (the remaining part of Jordan). The PLO moved from Jordan to Lebanon following the 1970 civil war, but since 1979 have resumed a relatively close relationship with Jordan. Recently King Hussain has had to incorporate representatives of increasingly popular Islamic groups in his government.

Lebanon, in some respects the most socially and politically Westernized state in the region in 1965, has proved the most turbulent in the following twenty-five years, due in the main to the vast number of religious groups competing for dominance, and to the involvement of outside elements: the Palestinians, Syria and Israel; and to a lesser extent Iran, Iraq and various Western nations. The Palestinians have played a major role in the state since 1970. Syria has been in control of parts of the country since 1976, with the defeat of General Aoun's Maronite (Christian) forces in 1990 the latest stage of their increasing dominance. Israel invaded Lebanon in 1978 and again in 1982, and remains in effective control of the South through a Lebanese militia.

A number of outside powers have played a distinctive role in the region. The UK and France had a historical colonial connection with the area, and continue to be important suppliers of arms, but in the period 1965–90 the USA and the USSR were the dominant powers. The USA sought to preserve its relationship with its Arab allies, notably the conservative monarchies of the Gulf, minimize Soviet influence, protect its oil interests. The USA has also provided Israel with considerable economic and military aid, and with diplomatic support in the United Nations. The USSR's principal regional allies have been Syria and Yemen PDR, but it has also supplied significant military assistance to Iraq. But the collapse of Soviet power has left the door open for increasing American influence, vividly highlighted by the phenomenal success of its military operation aginst Iraq in 1991. This does mean that the USA now has the best chance since 1948 to coax the states of the region into a settlement of the Arab–Israeli conflict. But as the problems of the Kurds and Shia's in Iraq demonstrates, there are plenty of other issues over which conflict may arise.

THE GEOGRAPHY OF IGNORANCE

This region is one of the most discussed and analysed of any on the globe, by external and internal experts. Its instability is obviously caused at one level by the coincidence of local social and cultural fragmentation, and great disparities in wealth, with the keen interest of outside powers in the source of that wealth, so important to the world economy. To secure their influence, the outside powers have connived at arming their proxies, and from time to time these arms are used when political pressures caused by the local fragmentation reach flash point.

There is considerable ignorance about how to develop these arid lands, about how to improve agriculture without damaging aquifers by overuse or by pollution, about how to improve urban services, particularly sanitation. But most of all there is ignorance over how to develop those political institutions which will allow the peoples of this area to live in harmony within each country and with neighbouring countries. New political institutions will have to reflect the fact that sovereignty is not an absolute concept, when so many cultural groups straddle arbitrary borders, when external sources of wealth (the oil-hungry markets of the developed world) simultaneously generate technological dependence. But they must also allow for the expression and realization of aspirations by the Islamic and Arabic peoples.

REFERENCES

Allan, J.A. (1985) 'Irrigated agriculture in the Middle East: the future', in Beaumont, P. and McLachlan, K.S. (eds) *Agricultural Development in the Middle East*, Chichester: John Wiley.

Al-Moosa, A.A. and McLachlan, K.S. (1985) *Immigrant Labour in Kuwait*, London: Croom Helm.

Beaumont, P., Blake, G. and Wagstaff, J. (1988) *The Middle East: A Geographical Study*, London: David Fulton.

Birks, J.S. and Sinclair, C.A. (1980) *International Migration and Development in the Arab Region*, Geneva: International Labour Office.

Blake, G. and Drysdale, A. (1985) *The Middle East and North Africa: A Political Geography*, New York: Oxford University Press.

Blake, G. and Lawless, R. (eds) (1980) *The Changing Middle Eastern City*, London: Croom Helm.

Blake, G., Dewdney, J. and Mitchell, J. (1987) *Cambridge Atlas of the Middle East and North Africa*, Cambridge: Cambridge University Press.

BP (1990) *Statistical Review of World Energy*, London: British Petroleum.

Clark, B.D. (1980) 'Urban planning: perspectives and problems', in Blake, G. and Lawless, R. (eds) *The Changing Middle Eastern City*, London: Croom Helm.

Clarke, J.I. (1980) 'Contemporary urban growth', in Blake, G. and Lawless, R. (eds) *The Changing Middle Eastern City*, London: Croom Helm.

Economist Intelligence Unit (1988) *Arabian Peninsula*, London: EIU.

Edge, S. (1990) 'Merger proves an attractive option', *Middle East Economic Digest* 19 January.

Encyclopaedia Britannica (1990) *Book of the Year*, Chicago.

Gischler, C. (1979) *Water Resources in the Arab Middle East and North Africa*, Wisbech: Menas.

Gould, St J. (1990) 'A review of the boundaries of Palestine and Israel with reference to a future Palestinian state', London (mimeo), Geopolitics and International Boundaries Research Centre, School of African and Oriental Studies.

Horowitz, D. (1975) *Israel's Concept of Defensible Borders*, Jerusalem: the Hebrew University.

Hourani, A. (ed.) (1988) *Cambridge Encyclopaedia of the Middle East*, Cambridge: Cambridge University Press.

Kemp, P. (1989a) 'Special Report: Saudi Arabia', *Middle East Economic Digest*, 31 March.

Kemp, P. (1989b) 'Investing in Islam', *Middle East Economic Digest* 21 April.

Kemp, P. (1990) 'Special Report: Saudi Arabia', *Middle East Economic Digest* 31 March.

Lawless, R. (1980) 'The future of historic centres: conservation or re-development', in Blake, G. and Lawless, R. (eds) *The Changing Middle Eastern City*, London: Croom Helm.

Lewcock, R. (1986) *The Old Walled City of Sana'a*, Paris: UNESCO.

Liden, A. (1984) *Security and Recognition: A Study of Change in Israel's Official Doctrine 1967–1974*, Lund: Studentlitteratur.

McLachlan, K.S. (1980) 'Development – the disaster of the oil boom', in *Middle East Review*, Saffron Walden: World of Information.

McLachlan, K.S. and McLachlan, A. (1989) *Oil and Development in the Gulf*, London: John Murray.

Middle East Economic Digest (1990) 16 November.

Middle East Review (1990), Saffron Walden: World of Information.

Morad, M. (1990) 'Towards a protection policy for the environment in the Arab World', *Arab Affairs* 1 (11): 96–106.

Naff, T. and Matson, R. (eds) (1984) *Water in the Middle East: Conflict or Cooperation?*, Boulder, Col.: Westview.

OPEC (1990) *Annual Statistical Bulletin*, Geneva: OPEC.

Parker, M. (1989) 'Come and join our club', *The Middle East*, April: 34.

Parker, M. (1990) 'The green revolution', *The Middle East* December: 52.

Rumaihi, M. (1986) *Beyond Oil: Unity and Develoment in the Gulf*, London: Al Saqi.

Schofield, R. (1986) *Evolution of the Shatt al'-Arab Boundary Dispute*, Wisbech: Menas.

United Nations (1988) *Demographic Yearbook 1988*, New York: United Nations.

Weinbaum, M. (1982) *Food, Development and Politics in the Middle East*, Boulder, Col.: Westview.

World Bank (1990) *World Development Report*, Oxford: Oxford University Press.

9

THE NON-ARAB MIDDLE EAST

Iran, Turkey and Israel

Keith McLachlan

INTRODUCTION: IRAN, TURKEY AND ISRAEL IN THE MIDDLE EAST CONTEXT

Iran and Turkey, the 'Northern Tier' States of the Middle East, stand, if not in stark contrast to, at least somewhat different from their Arab neighbours (Figure 9.1). Both states have (albeit imperfect) forms of democratic government. Both are grappling, so far with only modest success, with the establishment of new national institutions and state structures which transcend the defence of the existing political regimes. The consequences, among others, even in Iran and Turkey, are that the political geography of the area is characterized by insecurity, conflict and uncertain frontiers. In the following paragraphs the evolution of the Northern Tier States and Israel, the non-Arab component of the Middle East, in the period 1965–90 will be reviewed. The close linkages in oil and politics between them and the Arab countries should not, however, be overlooked at any stage.

The existence of oil reserves (BP 1990) accounting for more than 10 per cent of the world total, and oil exports at some 2 million barrels/day (EIU 1990), valued at $22 billion in 1980 and $16 billion in 1990, gives Iran a gloss of wealth (MEES 1990). Iran became increasingly an oil-based economy in the last quarter century as the proportion of national production from oil rose from some 11 per cent in 1965 to 40 per cent by 1975 and 21 per cent in 1985 through 1990, a depressed performance arising from the effects of revolution and war. However, in a real sense 'growth' generated by oil income is far from being 'development'. Oil revenues have enabled the country to import foreign goods and services but seem to hinder

213

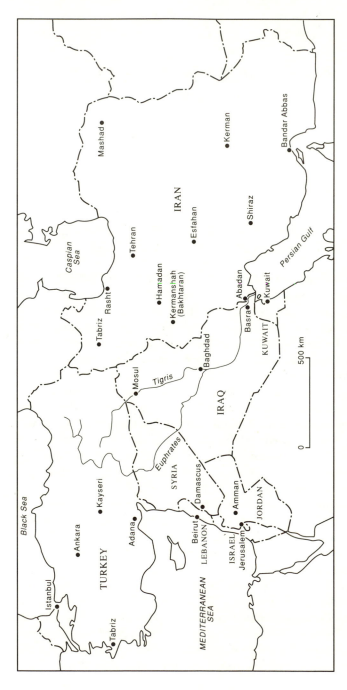

Figure 9.1 The non-Arab Middle East

improvement of domestic means of production from agriculture (McLachlan 1988) and industry (Stevens 1986), which has failed to compete with cheap imported goods.

Turkey, by comparison, is an oil importer. In 1987, 31 per cent of the value of exports went to the purchase of foreign oil (IBRD 1989). It has had to struggle to improve domestic output in order to be able to export goods to pay for fuel from abroad. Its performance in the key sectors of agriculture (3.3 per cent rate of growth in the 1980s) and industry (8.5 per cent) has been remarkably good particularly since the late 1970s (IBRD 1989). In that period Turkey transformed itself from an economically introverted country into a world-ranking exporter of manufactured goods, farm products and construction services.

Israel is a complex state, grafted into the Middle East region by political upheavals and wars in 1948, 1956, 1967 and 1973. Its territory was acquired by land purchase and conquest from the indigenous Arab population in processes which the latter never accepted. The original Israeli population was drawn from the Jewish diaspora in Eastern Europe, later augmented by Jewish refugees from the Arab states and elsewhere. The political structures within Israel are democratic but exclude from franchise and active involvement in government many of the large and rapidly increasing numbers of Arabs within the Israeli borders and in the occupied territories of Gholan, the West Bank and Gaza. Israel's boundaries have been constantly changing (see Figure 9.2) in response to war and to ideological claims for expansion to take in the entire area of Eretz Israel, the presumed territories of the pre-diaspora kingdom. Israel has changed its ethnic composition quite radically over the last twenty-five years. In 1987, for example, the population of Israel was 4.4 million. Since that time the immigration of large numbers of Jews from Ethiopia and latterly the USSR, from which up to 3 million Jews alone are expected to come to Israel, could almost double Israeli numbers, adding to pressures for territorial expansion.

Israel in a political sense has remained a constant within the Middle East scene throughout the period since 1948. The Arab countries and the Palestine Liberation Organization are determinedly hostile to the State of Israel and its supporters. Propaganda, terrorism and war have resulted from the deep antagonism between Israel and the Arab states. The fates of regimes and international policies in the region have been determined largely by this single factor in which Israel represents a major military power out of all

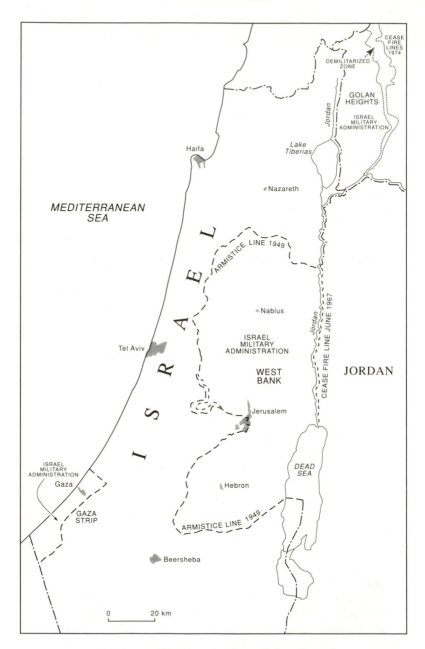

Figure 9.2 The changing boundaries of Israel

proportion to its population and territorial size and is surrounded by generally impotent neighbours none the less dedicated to its elimination. Thus far any attempt to resolve the problem through creation of a Palestinian state comprising Israeli lands occupied since 1967 has been thwarted mainly by Israeli intransigence on the issue.

DEMOGRAPHY

Iran and Turkey have the largest populations of the Middle East and together with Israel contain some 40 per cent of people within the twenty countries of the region (Beaumont et al. 1988). Iranian demographic growth came comparatively late, gathering appreciable momentum only after the mid-1950s. By 1990 the population was in course of an explosion in numbers and lay somewhere between 52 million and 55 million even taking into account a maximum loss in the Iran–Iraq War of between 750,000 to 1 million persons. Table 9.1 shows that Iran is a poor Third World country in its demographic characteristics despite the IBRD classification in the 'upper-middle-income' range.

Table 9.1   Demographic parameters in Iran, Turkey and Israel

	Population (mn) 1990	Growth rate 1980–8 annual %	Growth rate 1990 (%)	Population 0–14 years(%)	Crude birth rate/1,000	Life expectancy
Iran	56.1	3.0	3.1	44	41	63
IBRD upper-middle-income average			1.7	33	26	68
Turkey	51.7	2.3	2.0	35	30	64
Israel	4.6	1.7	1.7	32	22	76
OECD average			0.6	20	13	76

Source: IBRD (1990).

Turkey slowed in the years 1965–90 to a relatively moderate rate of population growth at less than 2 per cent per year though there were geographical variations between the more mature populations of the west and north and the youthful peasant populations of the centre and east, a polarization tempered only by steady migration to the Aegean and Black Sea coastal areas. The Turkish and in many ways the Israeli demographic situations differ from the rest of the Middle East. Whereas Iran and the Arab states of the Middle East see themselves as Third World countries and measure their populations against that standard, Turkey has aspirations to join the European Community and tends to use OECD averages as yardsticks. The figures in Table 9.1 demonstrate that Turkey has some way to go before its population structure begins to resemble the OECD states.

Israel, meanwhile, offers a complicated picture in which there is an assortment of early and mature structures coexisting within discrete parts of the community correlating with the source, period and age of migrants to the territory. There is also a major discontinuity in population regime between the Jews and the Israeli Arabs. The latter exhibit the norms of other Arab communities of the Middle East, with high birth rates and large average family size, though also responding in a qualitative way to the high standard of medical and social services available within the Israeli state. Taking into account the Arabs in the occupied territories, the higher growth rates of population among the Arabs than the Jews has given rise to fears that the Jewish cultural nature of Israel will be eclipsed as Arabs eventually outnumber Jews (Soffer 1989).

Table 9.2 Urbanization in Iran, Turkey and Israel

	Urban population 1965	1988	Growth rate 1965–80	Main city (%) 1980	Cities over 500,000
Iran	37	54	5.1	28	6
Turkey	34	47	4.2	24	4
Israel	81	91	3.5	35	1

Source: IBRD (1990).

All three states have undergone a revolution in recent years in the distribution of population. Rapid urbanization has occurred both from natural growth processes within the urban communities and as rural migrants have flooded into the towns and cities (Table 9.2). In Israel there was an ideological bias towards rural communal development during the initial state-building period. Sophisticated co-operative organizations such as kibbutzim and moshavim were established to construct a productive agricultural base for immigrant Jews and tie the settlers to the land. This commitment lost its force in recent decades as Jewish labour on the land was replaced by itinerant Arabs and the mode of housing immigrants changed from rural commune to new town. Greater Tel Aviv area dominates the settlement hierarchy with almost 30 per cent of the urban population. All other towns, including Jerusalem, being comparatively insignificant.

Iranian urbanization in recent decades has accelerated to become among the most rapid in the world outside Africa. A combination of a thorough-going land reform, rapid economic growth in the second half of the 1960s and, above all, the unrestrained oil boom of the mid-1970s led to an explosion in urban population and a rise in the flow of emigrants to the towns. Iranian towns and cities were free of major shanties and there was a perceptible quality to urban planning through provision of public utilities, broad streets and leisure areas in the period before 1980 and the onset of the eight-year war against Iraq.

In Turkey urbanization favoured the west of the country, especially Istanbul and Izmir, though Ankara in the centre and Alexandretta in the south have also been centres of growth. Much development in urban areas was very poor quality shanties of the so-called *gecekondu* type, literally built overnight to exploit squatters' rights. Large areas of Turkey's cities are blighted with shanties lacking drainage, electricity and other basic amenities. The flight from the land in the Kurdish areas of the east as a result of military insecurity (a problem longstanding before the Kurdish refugee crisis of 1991) has added to the problem in recent years. As measured against EC standards, Turkish cities lag severely behind and even compare badly with average Iranian levels. Ankara, with a population of more than 5 million, suffers badly from pollution, and only Istanbul with its intensive commercial activity and legacy of splendid Islamic and Ottoman imperial architecture stands out as attractive in an otherwise poor but burgeoning urban environment.

Urbanization in Israel is advanced both in the proportion of population in the towns and cities and the quality of the urban areas. Controversy surrounds much of the planning of the massive urban construction in Israel since its foundation in 1948. Political and military considerations have constrained the authorities to ensure a wide geographical spread of population through the building of new towns and the expansion of older villages and small towns. At the same time, the maintenance of open frontiers to new Jewish settlers from abroad has demanded a hectic pace of development of new urban settlements simply to provide basic housing for migrants and an expanding indigenous population. It is projected, for example, that Israel will have to house as many as 2 million or 3 million immigrants from the USSR over the five years to 1995.

Israel has taken on the protection of the great city of Jerusalem, though without the blessing of either the Arabs or the international community. Jerusalem, largely an Arab city before 1948, has been made into the capital of the state of Israel. Jewish areas of worship such as the wailing wall have been given priority for access and maintenance over Islamic centres such as the Al-Aqsa mosque and Christian Churches traditionally associated with the site. Old Jerusalem has been cleared in parts to make way for high quality housing for newcomers, while the outer rim of the basin in which the city sits has been taken over by large Jewish housing complexes, altering the old skyline. Within this internationally important city the Arab population is disaffected and tensions run high on all sides. Israeli Arabs remain in their established small town and village settlements, maintaining in urban form a cultural separation from the Jews surrounding them.

ECONOMIC CHANGE 1965–90

The economic fortunes of the three states have been mixed. In all cases, however, the pace of economic change has been profoundly affected by variations in the price of oil. Naturally, Iran, as an oil-based economy, reflects this most clearly. Iran was the first major petroleum exporting country of the Persian Gulf region producing its commercial oil in 1908. The oil shock of 1973–4 appeared to benefit Iran but demand for oil fell thereafter and the industrialized nations systematically reduced their imports of crude by finding more efficient ways of using energy and diversifying their sources of both petroleum and other energy sources. The balance between supply and demand for oil increasingly favoured importers and from the mid-1970s Iran was producing 6.5 million barrels per day (b/d), second only to Saudi Arabia in the Persian Gulf region. The oil shock of 1973–4 was advantageous for Iran but demand for oil fell thereafter and the industrialized nations systematically reduced their imports of crude by finding more efficient ways of using energy and diversifying their sources of both petroleum and other energy sources. The balance between supply and demand for oil increasingly favoured importers from the mid-1970s except for brief periods of international instability such as the Iranian revolution of 1979, the Iran–Iraq War of 1980–1 and the Gulf crisis of 1990–1.

Iranian output was affected by the Islamic revolution. New policies and the upset of revolutionary activities within National Iranian Oil Company dictated that output be reduced below 3 million b/d. The Iraqi invasion of Iran in September 1980 brought protracted warfare and considerable damage to Iranian oil production and export facilities occasioned both by air raids and a run down of care and maintenance. Once the war with Iraq ended in August 1988 there was a slow improvement in Iranian output and exports again rose above 2 million b/d. Given Iranian reliance on oil, which accounted for more than 90 per cent of the value of exports, 65 per cent of government revenues and 20 per cent of GDP, the changing volumes of exports meant crucial changes in economic circumstances. The oil price was the other key variable, trends for which, other than for the short term effects of regional crises, ran against Iranian interests. Oil income rose above $20,000 million each year after 1973 but disappointed expectations thereafter, stagnating or falling, with a particularly lean period in the 1980s when in some years revenues were less than half the level of the 1970s (Table 9.3).

Table 9.3   Iranian oil industry 1965–90

	Output ('000 barrels/day)	Revenue (US$ mn)
1965	1,908	514
1970	3,845	1,109
1975	5,385	19,634
1980	1,480	13,286
1985	2,215	13,255
1990	3,125(estd)	21,000

Sources: BP (1990); OAPEC, *Annual Reports* (various years); MEES (1990); EIU (1990).

Economic development in Iran was fast and positive in the period 1964 to 1974. Oil revenues were used as a means of funding a productive programme of industrialization in which the state set up large-scale projects in iron and steel at Esfahan, petrochemicals on the Persian Gulf coast and machine building at Arak in central Iran. The private sector contributed automobiles, textiles, leather goods and other light industries. Overall a rate of growth in real terms of

12–14 per cent annually was attained in that period. A specific regional industrial development policy was pursued, bringing small industrial estates to as many as thirty provincial towns and cities.

Agriculture was a less comfortable experience for the authorities. Land reform began well under the Shah as a means of establishing a peasant farmer regime but lost its way as the government unsuccessfully attempted to achieve modernization of farming on large-scale commercial units under central control. Farmer confidence evaporated and, under the additional stress of rapid expansion of the urban economy as a result of the oil boom of the 1970s, migration away from the countryside grew, investment in farming stagnated and the new farm structures failed to establish themselves as viable concerns. There was a great push to develop much-needed water resources by constructing dams on the country's great river systems *inter alia* at Aras, Karaj, Dez, Manjil, Zayanderud and Lar. Much of the new water was channelled to use in urban water supply and hydro-electric power. Little more than 100,000 net hectares of virgin land was brought under the plough through these irrigation developments. Elsewhere there was some move into well pumping but this was often at the expense of damaging the traditional *qanat* water provision systems (McLachlan 1988). By 1979 Iranian agriculture appeared comparatively neglected. Uncertainties of rainfall had not been compensated for by improved irrigation. Food imports rose dramatically during the 1970s to $2,000 billion or 10 per cent of total imports by value annually and the country ceased to be self-sufficient in agricultural produce even in years of good rainfall.

The rapid economic growth of the period 1964–74, averaging well over 10 per cent each year in real terms, changed the geography of Iran. The Persian Gulf coastlands, formerly little more than an alien, oil-producing enclave, were integrated firmly into the national structure as the iron and steel, petrochemical, transport and military sectors were greatly expanded from the Khuzestan plain through to the new principal national port at Bandar Abbas and the naval station at Chah Bahar. Industry, previously concentrated almost exclusively in Tehran, penetrated to most large towns. Oil and natural gas were made reliably and cheaply available throughout the country by construction of a series of pipelines. At the same time, population moved to the growing cities, especially changing the urban face of the south. Economic strength meant that Iran's military power was much in evidence as the Shah increased spending on defence after 1973. The state became the dominant regional influence in the Persian Gulf area with an expanding role both in the Northern Tier in association with Turkey and Pakistan and in the Indian Ocean.

Turkey's economic performance in recent decades shows, in comparison with Iran, that the oil-importers can and often do develop as or more successfully than the oil-exporters (see Table 9.4). While Turkey did suffer recession and shortages from the rising cost of oil in the 1970s, it ultimately reformed its economy and returned a growth rate for 1965–80 at 6.3 per cent annually, slightly in excess of those achieved in Iran in the same period (6.2 per cent). The rate of economic change in Turkey was no less erratic than that in Iran but the qualitative development of the country's economic structure was much more convincing. Growth was spread across the economy with high rates of change in agriculture accelerating consistently and perceptibly in the 1980s. Manufacturing was even more remarkable as the value of output increased to peak in the mid-1980s as Turkey became a major supplier to the warring states of Iraq and Iran, partly in return for oil. Trade dependence on Iraq and Iran created difficulties in the late 1980s as they were overwhelmed by foreign debts, falling oil incomes and a diminished ability to import goods and services.

Table 9.4 Rates of economic change in Iran, Turkey and Israel (per cent: annual average)

	Agriculture		Manufacturing		Services		Total GDP	
	1965–80	1980–8	1965–80	1980–8	1965–80	1980–8	1965–80	1980–8
Iran	4.5	2.0	10.0	1.0	13.6	7.0	6.2	1.5
Turkey	3.2	3.6	7.5	7.9	7.6	5.1	6.3	5.3
Israel	—a	—a	—a	—a	—a	—a	6.8	3.2

Source: IBRD (1990).
Note: a Data unavailable.

Turkish agriculture was an impressive performer, providing adequate food to feed the nation in all but drought years. Output from farming was greater in Turkey than in all other Middle Eastern countries together and was a source of foodstuffs for many of the regional states. Mechanization went ahead rapidly as population left the land, notably in the large private estates of the east and output per hectare and per person rose dramatically. Considerable financial resources were put into the development of dams in Anatolia, including the existing Keban and Ataturk reservoirs. Other projects on the Euphrates and Tigris rivers were in hand to increase the area under irrigation by 1.6 million hectares in addition to providing important hydro-electric power generation capacity.

Modern industrialization began early in Turkey with mining and the construction of heavy industry but these basics were not developed, and industrial growth came slowly before and after the Second World War until the 1970s when import substitution and assembly industries were established in some number. Although patchy in sector coverage and not entirely efficient, the growth of light industry enabled a large-scale expansion of manufacturing in the 1980s in which Turkey became an exporter of finished goods.

The elaboration of the urban industrial base of the state was underpinned by new banking and service sectors, some rather inexperienced and fragile. With particular effect in the 1980s, international tourism grew as a provider of foreign exchange and employment, worth more than $1 billion per year. This development was needed to compensate for a fall-off in remittances sent back to Turkey by workers overseas as European states such as the Federal Republic of Germany reduced their demand for guest workers and as the Arab economies stagnated at the end of the oil boom of 1979. Despite the recession, Turkish workers abroad remitted a net $1.76 billion in 1988 (IBRD 1989), equivalent to 45 per cent of the country's foreign exchange reserves in that year. Thus, Turkey was a far more diversified economy by the end of the 1980s than it had been twenty-five years earlier. Though not yet balanced and mature, there were signs that Turkey had successfully launched itself on to a path of sustainable development, making the best use of indigenous natural resources and able to survive as a competitive exporter. Perhaps alone of the countries of the Middle East and North Africa, Turkey was not economically a rentier state though it continued to share one characteristic with its neighbours – political uncertainty.

The Israeli economy is atypical for the region and is also very different from the OECD-style economies with which it is normally classified. It must always be taken into account that Israel is as much a religious/ideological expression of intent by non-resident Jews as a material economy. Overseas development assistance averaging $1.5 billion, military aid running at up to $3 billion and private charity transfers of funds averaging annually over $1 billion play a dominant role in maintaining Israel. At the same time there are political considerations which bear heavily on the economy. Israel has a very unusual position in accepting all bona fide Jewish migrants from overseas as a matter of principle despite the enormous costs that such a policy incurs. The country is surrounded by declared military enemies with large populations and in some cases formidable military establishments. This has demanded constant vigilance on matters of security, a citizen army kept permanently on call and a large standing defence force with advanced technology at its disposal to enable defence of a tiny surface area against surprise attack. In most years Israeli defence expenditure runs at more than 20 per cent of GDP (International Institute for Strategic Studies 1985).

Table 9.5   Changes in economic structure in Iran, Turkey and Israel

| | *Agriculture* | | *Manufacturing* | | *Other industries* | | *Service* | | *GDP 1988* |
	1965	*1988*	*1965*	*1988*	*1965*	*1988*	*1965*	*1988*	*($ bn)*
Iran	26	23	12	13	24	21	38	43	105.0
Turkey	34	17	16	27	9	10	41	46	64.4
Israel	—a	5	—a	22	—a	28	—a	45	29.8

Sources: IBRD (1990); Central Bank of Iran, *Annual Reports and Balance Sheets* (various years).
Note: a Data unavailable.

The structure of the Israel economy is mature in comparison with other states of the region. Agriculture, though effective and productive, produced only 5 per cent of GDP, while manufacturing was 22 per cent (Table 9.5). Growth of the economy has been consistently rapid, albeit less impressive during the 1980s. Israeli agriculture is a model for regional development, offering intensive use of irrigation water to produce export crops of oranges, early fruits and flowers accounting for 5.5 per cent (1988) of all exports. Industry is well developed. In addition to traditional items such as diamonds (29.0 per cent), important amounts of machinery, including defence equipment (29 per cent) and chemicals (22 per cent) are exported.

ENVIRONMENTAL ISSUES

Pressure on natural resources has been acute for many years in the region as a whole. Environmental degradation was a recognized problem in Turkey and Iran long before consciousness of green issues developed in the industrialized world. Deforestation, severe soil erosion and air pollution, the latter especially in parts of Turkey, made themselves felt as early as the First World War when much of Turkey's Anatolian forests were cleared as part of the war effort. These and other damage to the environment were worsened by growing population, cultivation of physically marginal areas, extension of uncontrolled tractor farming and demands for raw materials to support industrial development. The very high rates of urbanization in Turkey and Iran have however contributed to the abandonment of rural farmland, and the neglect of old terraces. This may have both postive and negative aspects, depending on the degree of previous overuse of the land and its potential for instability.

The shock to the environment in recent decades through the hand of humans could aptly be described as 'self-destructive growth' (Bowen-Jones 1983) in so far as losses of top soil, non-regenerating woodlands and overgrazing of natural rangelands brought irreversible damage. Attempts to limit damage are best seen in Turkey where controls were put on grazing in forest areas before 1965, though not effectively in the Tauris Mountains where grazing and indeed crude ploughing led to widespread loss of soil (Brice 1966). In the west of the country land reform helped to restrain the pace of land deterioration. Land reform through the Topraksu organization brought not only land consolidation and distribution but also a complete system of economic, social and above all environmental improvements, including full watershed management, reafforestation and education in tractor ploughing techniques. Unfortunately the land reform went ahead slowly and never affected the badly blighted landscapes of the centre and east.

For historical reasons, largely tied up with the nature of land ownership, the late and limited arrival of mechanized farming and the fear of government punishment for offenders found removing trees in the state domain, Iran suffered less extensive damage to soils and forests than Turkey. None the less, the margins of the large forests in the Caspian-Elburz region of the north and the vestigial forests of the Zagros in the south have been attacked by farmers and even one or two state agricultural projects. In the 1960s and 1970s the government gradually implemented scientific management of the forest and to a lesser extent rangelands. Some reafforestation was undertaken especially around Tehran, though much of this work was for cosmetic and aesthetic reasons. A decline in farming in marginal areas both on the rims of the great deserts or at higher altitudes in the mountain chains took some pressure off the more fragile areas of the country. The withdrawal from the more isolated and difficult zones of cultivation arose mainly from the side-effects of oil-based developments in the towns with effect from 1964 though most markedly after 1973. Land reform in stages two and three, which occurred in the period beginning 1964, was inimical to small peasant farming and brought a widespread atmosphere of rural insecurity which in turn accelerated the abandonment of marginal farmlands.

The Iranian oil industry inevitably creates special problems for the environment. Other than the visible marking of the landscape in the oil-field areas and the air pollution created by the existence of oil refineries, the environmental impact the industry has thus far been well controlled. This situation changed when major lapses occurred during the Iran–Iraq war (1980–8). A series of air and missile attacks against oil export terminals, offshore oil-field installations by both sides in the war, though particularly against Iran by the Iraqis resulted in severe levels of oil leakage into the waters of the Persian Gulf. A regional organization was set up to research the effects of oil pollution but made little real progress.

Israel has a strong record of environmental management. Its involvement in reafforestation of large areas of the country and application of scientific management of land and environment is matched nowhere else in the Middle East. It is not without its problems, however. The very density and rapidity of urban population growth pose in recent years special difficulties as good quality soil and climate areas are built over. Conservation of the landscape in the Holy Land has been almost entirely ignored as is indicated by the permission given for the vast and generally ugly housing areas around the city of Jerusalem. Security considerations have meant the construction of Israeli settlements in border regions in areas of outstanding international natural interest such as the Dead Sea basin, while tourist facilities and port developments have visually scarred and brought sea pollution to the area around Eilat on the Gulf of Aqaba.

Perhaps the most worrying environmental issue in the Middle East region as a whole is that of depletion of underground water stocks and competition for use of surface waters. The water supply problem has become critical for many states of the region in the last twenty-five years. The very processes of economic development and population growth which have so characterized the period since 1965 have been responsible for pushing a difficult water supply situation into crisis and into a growing but little ventilated appreciation that water is already an intractable limitation on economic growth. Israel is in the most acute position. It relies on water derived from outside its national borders for at least a quarter of its supply. Water tables everywhere within Israel are depleted almost to exhaustion. Water used for agriculture is increasingly brackish and, though managed carefully, presents the risk of permanent damage to the soil. Urban supply can now be provided for only with a resort to desalination of sea water at great expense. The worst situation is observed in the occupied territories, notably Gaza, where the availability of water per head of population is now declining parlously. Turkey has least problems with water supply thanks to the new storage and future potential for reservoir storage in the great river systems of the Tigris and Euphrates. The exploitation of these waters exports the water problem to the downstream states. Iran, living with a late arrived and apparently uncontrollable boom in population numbers, has moved into the first phase of its water crisis. Water tables are falling, urban water provision is increasingly expensive and, for Tehran, even uncertain. The benefits of dam building in the 1960s and 1970s are being lost as on the one hand the easily developed reservoir sites are used up and on the other hand existing dams suffer from silting and could cease to be major suppliers within twenty to forty years. Iran for some time will be able to use its wealth in petroleum to provide water for agriculture, industry and urban consumption without the necessity for major environmental damage. But water costs will rise disproportionately and add to Iran's competitive disadvantages once oil is no longer a principal earner of foreign exchange income.

THE ISLAMIC REVIVAL: SOCIAL AND POLITICAL CHANGE

The most thorough-going change within the region in recent years was the Islamic revolution of 1979 in Iran. It signalled a successful challenge to the industrial and ideological paramouncy of the West and a serious revival of Shi'ite philosophies in competition with orthodox Islam. The outstanding manifestation of the revolution was geographical since the Shi'ites are concentrated in Iran (48

228

million), southern Iraq (8 million) eastern Turkey (1.6 million) and Lebanon (1.1 million) with a scattering elsewhere in the Persian Gulf region (0.9 million). The basically quiescent phase of Shi'ite Islam ended with the emergence of Ayatollah Khomeini as the accepted single interpreter and arbiter within the non-orthodox Islamic world. His writings and teachings proposed Islam as a total way of life in which there was no differentiation between religious and secular authorities, Muslims should actively cleanse Islam of corruption and heresy and the non-Islamic 'satans' could be confronted. The change from quiesence to aggression was signalled by the claims that the time had come for the emergence of the 'hidden' imam from occultation. Ayatollah Khomeini was regarded at the time as being the na'eb-e imam or the representative of the hidden imam. So significant a turning point in history for the Shi'i created conditions for a holy war (jihad), for the sweeping away of corrupt secular heads of state in Islamic countries and for self-assertion where the Shi'i were politically oppressed and religiously persecuted as in Lebanon.

In Iran the social upheaval of the revolution altered the structure of society. The ruling elite hazar (one thousand) families were all but eliminated through exile and execution. Social morality reverted to strict Muslim rules, reversing earlier reforms such as female emancipation and suffrage, application of a civil code of laws and freedom of expression. Similar changes occurred elsewhere in the Shi'ite communities of Turkey. The rise of Islamic feeling deeply affected Turkey, including both supporters of Khomeini and Sunnis reacting defensively to Khomeiniism by radicalizing orthodox Islam. So great was the rise of Islamic reassertion in Turkey that foreign commentators took the view at the end of the 1980s that socially and politically the country was facing the most serious attack from Islam since the foundation of the republic (Mackenzie 1987).

Even Israel was not left untouched by the forces of religious revivalism. Extreme orthodox Jews increased their attacks on the secular aspects of the Israeli state. Given the balancing role played by a number of orthodox deputies in the Knesset (parliament) in the complicated Israeli democratic system of coalition government, the fundamentalist sects of Judaism were well represented as a political force and able to get their ideas implemented despite their minority position and the dislike of them by the mass of Israelis.

In addition to the tremors induced into society as a result of

Islamic revivalism and the reactions to it, politics was profoundly affected. Ayatollah Khomeini devised entirely new concepts of rule and constitution for Iran, assuming always that his ideas would be universally applicable. Governance was to be the direct rule of the men of god manifest in a *velayat-e faqih*, the guardianship of the Islamic jurist. He was at one and the same time the arbiter for the state and the source of emulation *marja al-taqlid* for every Muslim. These concepts were enshrined in the Iranian constitution of 1980. Ayatollah Khomeini had a further nine years to entrench them before his death in 1989. The process of succession was rapid but did mean a devaluation of Khomeini's ideas since no one of his religious and political stature was available to succeed him. The prosecution of the inconclusive war against Iraq after 1982 diminished the perception that Khomeini was really the forerunner of the coming of the hidden imam. This assisted a gradual reduction in the aggressive and revivalist philosophy of the Iranian revolution after his death and the ideas which threatened at one time to spill across the Middle East and the entire Islamic world lost strength in Iran though they persisted as powerful forces elsewhere.

Turkey mainly resisted the Islamic revivalism prompted by the Iranian revolution but faced continuing political threat from Islamic movements of an indigenous variety. The coalition governments in Turkey preceding the army *coup d'état* of September 1980 were feeble and content to ally with Islamic movements where they offered a means of inhibiting the activities of the extreme left wing. After the army coup in 1980 Sunni Islam was co-opted once again but more formally by officially supporting the teaching of Islam for example, to act as a counterbalance to alien ideologies of the left and Khomeiniist Islam. In consequence the influence of Islam in Turkish political life grew apace in the 1980s. Attempts by the Motherland Party under Turgut Ozal, which ruled the country from 1983, to curb the conservative Islamic forces towards the end of the decade tended to exacerbate social and political divisions on the question of the role of Islam in Ataturk's secular state and the appropriateness of Turkey seeking membership of the European Community.

In Israel, political change has been slow within the Jewish electorate. For Israeli Arabs life has been extremely difficult. They remain outside the ambit of organized public political life yet are unable to play a full role in the main Palestinian freedom movement, *Intifada*, which above all affected the occupied territories beginning in Gaza in December 1987. *Intifada* brought violent but unarmed street fighting by Palestinians against the Israeli authorities. Many deaths resulted, most of them young Arabs killed by Israeli bullets. The *Intifada* movement profoundly shocked the Israeli nation since it was both widespread and persistent and brought out a level of violence among the Israelis themselves which undermined a confidence in the moral position of the state both at home and abroad. Meanwhile, the PLO developed a peace process in the international community to press for a settlement of the Palestine issue within the context of United Nations resolutions, essentially demanding a return of Israel to its 1967 boundaries (GRC 1990). Israeli intransigence on the issue of treatment of Palestinians in the occupied territories and their refusal to meet and work out a solution with the official representatives of the Palestinians, the PLO, changed attitudes towards Israel in Europe and the USA. By 1990, Israel was increasingly isolated internationally and its moral position *vis-à-vis* the Arabs deeply undermined. In the event, Israel was rescued from its deep embarrassment, though only temporarily, by the Iraqi invasion of Kuwait in mid-1990 and the support by the PLO Leader Yasser Arafat for Iraq. But it remained that one of the most important changes in the political geography of the region, the creation of a Palestinian state, became far more a possibility by 1991 that it had in 1965.

INTERNATIONAL CONTEXT FOR IRAN, TURKEY AND ISRAEL

The Iranian revolution brought a deep discontinuity to the geo-political configuration of the region. Iran was groomed by the Western powers to take over the British role as peace-keeper in the Persian Gulf after the withdrawal of UK forces in 1971. It fulfilled its given task well, helping to put down revolt in Oman, excluding Iraq, an ally of the USSR, from the Gulf and fostering conservative forces in the Arabian peninsula. As a member of the Central Treaty Organization, Iran was an important sector of Western containment of the USSR in that period. In 1979 the revolutionary Islamic republic revoked all its links with the West in oil, economy, defence and diplomacy. From being seen as an island of stability sympathetic to the West, Iran became alien and hostile. In reality, Iran turned to ideas already well embedded in its political culture and antipathy to all foreign influences, particularly those of the major powers whether of the West or East. The revolution was ultimately inward looking, concerned with aspects of Shi'ite divinity and domestic morality together with the domestic consolidation of its power. Unfortunately, it was not realized that rhetoric against the West and other states was largely a matter for local consumption. Even the actions against the US embassy and foreign oil interests in 1979 and 1980 were above all designed to establish Iranian sovereignty rather than redefine the world system. Iranian threats against the outside world and an extremely aggressive set of policies were interpreted internationally however as evidence of Iran acting as a hostile and destabilizing force. The country became entirely isolated except for links with Syria and to a lesser extent Libya.

In September 1980 Iraqi forces attacked Iranian border territory in Khuzestan, Kurdestan and Lurestan. The geography of the war was highly localized for much of the time to a narrow strip of land between the two countries in which the land war was fought to a stalemate in 1988. From 1985 fighting spread first to aerial attacks on civilian and economic targets using aircraft and rockets and second across the waters of the Persian Gulf. Iran was compelled to change the lines of oil export, retreating gradually from Bandar Khomeini and Kharg Island to Sirri and Larak at the mouth of the Gulf. Large displacements of population took place, with more than a million Iranians fleeing from the war zone to central areas of the country. Iranian efforts after 1982 to take the war into Iraq and to assist in the liberation of the Shi'ite communities of Iraq failed, and the threatened changes in regional boundaries never materialized.

Iraq was also the cause of the second regional upset, the invasion of Kuwait in 1990. Iranian geopolitical interests were adversely affected by the Iraq move and the later possibility of a break-up in Iraq in which Iran wished its interests to be taken into account. The Iraqi invasion did a great disfavour to Iran and other nation states struggling to assert their sovereignty in the region by bringing into the Persian Gulf the armed might of the USA and its allies as the prevailing force. In the course of the short period of twenty years the local states had been given the opportunity to manage local strategic affairs between themselves and lost it.

Turkey played a key role as a supplier to Iran and Iraq during the 1980–8 war, for many years providing the only oil outlet for Iraqi crude exports via a pipeline to the Mediterranean. The outbreak of the Gulf Crisis of 1990–1 saw Turkey as a NATO ally of the USA and prospective European state closing down this very pipeline and sealing the Iraqi border against all imports and exports other than those permitted by the UN. Turkey, like Iran, kept a secondary watching brief on the military situation in Iraq so that it could intervene should its national self-interest be affected by the hostilities.

THE GEOGRAPHY OF IGNORANCE

The contrast between Israel and the Northern Tier States is at its greatest in matters of freedom of information. Israel is well documented from earliest times. Excellent geographies of the Holy Land abound (cf. George Adam Smith 1896; Charles Foster Kent 1920). The creation of the State of Israel and its political consequences have given rise to a vast outpouring of printed matter. In geography, Yehudah Karmon and his many successors and students have established a very wide and analytical literature in Hebrew and other languages inquiring into all aspects of the discipline. At the same time, the ease of access to Israel and the ability to foreigners to live and work in agricultural settlements in Israel adds to the range of contacts to dispel 'ignorance'.

Turkey has only recently made an effort to open itself to comparatively unrestricted domestic research and survey by foreign scholars or travellers. The statistical base is also rapidly improving in breadth and accuracy. Technical advances in remote sensing and topographic survey have also been consolidated. Given the enormous size of the country, however, there is much research yet to do in all aspects of geography from land use and ownership to the nature of recent urbanizations. Eastern Turkey has special difficulties, with a growing security problem because of Kurdish and Islamic unrest, thus travel and safe access are not always guaranteed.

Iran stands in a sense apart as a geographical focus for research and even basic description. Insecurity, wars and closed frontiers have kept much of Iran out of reach of scholars for long periods in the present century. But in a brief flowering of mainly unrestricted research between 1965 and 1978 many Iranian students and scholars made their mark both in Iran and abroad. Travel passes were abolished for foreigners and close surveillance was partially relaxed at that time. Since 1979 revolution and war have encouraged official suspicion of foreign research of all kinds and limited resources have held back Iranian scholars. In the decade 1980–90 serious publications were based on work in the field done earlier and much of the rest was no more than speculation from a distance. There is an immense amount of research to be done to make up for lost time, though changes will be needed first in the attitude of foreign scholars to accommodate the now rigorously Islamic and independently minded Iranian society.

SUMMARY

The geography of the Northern Tier States and Israel since 1965 has changed continually. In economy, Israel and to an extent Turkey have opened up to become important by scale and value while structurally moving towards greater maturity. Geographical distributions of wealth and population within the three states have altered, though income per head has advanced less dramatically for most of the people than the figures suggest. Israel's population is by far the wealthiest with $8,650 per capita against $2,500 in Iran and $1,280 in Turkey. More radical changes are apparent in the numbers and distribution of population. The region has begun a population explosion which will entirely alter its economic and political prospects; Iran is anchored firmly as a current oil-based economy with a difficult and uncertain future beyond the oil era but making it the most densely settled state of the Middle East region. The political geography of the non-Arab countries has become infinitely more complex in recent years. In 1965 all the three states were clearly in the Western camp. By 1990 Iran was at odds with the USA and much of the rest of the world. Turkey was an ally of the USA and was identifying itself economically and politically with the European Community. Israel stood in an insecure linkage with the USA on a basis that seemed to be becoming increasingly tenuous. Internally, the states faced great strains by the end of the review period. Revolutionary Islam had overturned social structures and basic attitudes in Iran. *Intifada* and the partial international legitimization of the PLO altered irrevocably Israeli moral superiority on the Palestine issue and brought the first possibility of a UN-backed drive for a final solution of the long standing conflict. The outlook for this area at the end of the twentieth century is above all uncertain and religious revivalism points to continuing political turmoil.

REFERENCES

Beaumont, P. (1985) 'The agricultural environment: an over-view', in Beaumont, P. and McLachlan, K.S. (eds) *Agricultural Development in the Middle East*, London: Tauris.

Beaumont, P., Blake, G. and Wagstaff, M. (1988) *The Middle East*, London: Fulton.

Bowen-Jones, H. (1983) *Agriculture in the Arabian Peninsula*, London: Economist Intelligence Unit.

BP (1990) *BP Statistical Review of World Energy*, London: British Petroleum.

Brice, W.C. (1966) *South-West Asia*, London: University of London Press.

CBI (1986) *Annual Report*, Tehran: Central Bank of Iran.

GRC (1990) 'The Palestine State', Geo-politics and International Boundaries Research Centre, School of Oriental and African Studies, verbal communication by PLO Representative.

Economist Intelligence Unit (EIU) (1985) *Regional Review: The Middle East and North Africa*, London: EIU.

EIU (1990) *Iran: Country Report*, no. 4, 1990, London: EIU 4: 15–16.

IBRD (1989) *World Development Report*, Oxford: Oxford University Press.

IBRD (1990) *World Development Report*, Oxford: Oxford University Press.

International Institute for Strategic Studies (1985) *The Military Balance*, London: IISS.

Kent, C.A. (1920) *Biblical Geography and History*, New York: Scribners.

Mackenzie, K. (1987) Leading article, *Observer*, 18 January.

McLachlan, K.S. (1988) *The Neglected Garden: The Politics and Ecology of Agriculture in Iran*, London: Tauris.

McLachlan, K.S. (ed.) (1991) *Sovereignty, Territoriality and International Boundaries in South Asia, the Middle East and North Africa*, London: School of Oriental and African Studies.

McLachlan, A. and McLachlan, K.S. (1989) *Oil and Development in the Gulf*, London: Murray.

MEES (1990) *Middle East Economic Survey*, Nicosia, 34 (12/13): A1–A5.

Smith, G.A. (1896) *The Historical Geography of the Holy Land*, London: Hodden & Stoughton.

Soffer, A. (1989) 'The demographic timebomb in Israel and the occupied territories', Tel Aviv (private communication).

Stevens, P. (1986) 'The impact of oil on the role of the state in economic development – a case study of the Arab world', paper delivered to Brismes Conference, School of Oriental and African Studies, London.

10

CONCLUSIONS
A growing sense of urgency
Graham Chapman and Kathleen Baker

Harold Brookfield (1975: ix) wrote sixteen years ago that 'the terms "developed" and "underdeveloped" occur within quotation marks wherever they appear, for the fundamental argument is that there is only development, but development of different qualitative orders.' Brookfield's point was twofold. First, that the process of development in the rich countries of the North 'is part of the same process of development of Africa or Asia towards a dependent, skewed economy, which we call "underdeveloped"', and second that development had only one meaning: change. The connotation that change, in the sense of increasing material consumption, is inevitably progressive, was strong when Brookfield was writing. Perhaps it is less so now that in the North there is a growing awareness of the ecological price paid for that level of consumption. So even that 'development' may not be 'development', simply change.

Geographies traditionally have been written about countries: it is what the word means – a description of part of the geo. The explanation for most of the observed patterns in a country has been in terms of correlations between variables at that place. Climate and cultural practice explain cropping patterns and preferences, geology and river transport the exploitation of resources. Even export crops and ports are explained in terms of an opportunity seized locally to benefit from an export opportunity. An opposite approach following the arguments of Furtado, Frank or Prebisch would stress that right down to the last and furthest peasant, explanation is not to be sought locally, but must be traced back to the impact, not to say machinations, of the metropole of world capitalism. All the geography of a place, except its basic environment, is explicable in terms of external dependency.

The first and most immediate impact that the preceding chapters

236

have had on us, their editors, is to reveal the way in which issues of dependency have crept into description and explanation at every stage: although not necessarily overtly, and not necessarily concluding one way or the other that there is a sole explanatory force. The opening introduction used material that looked forward to the succeeding twenty-five years, which for many countries would be their first experiments with independent political authority. Doubts about the future expressed by external observers were couched in internal terms of what Africans were and had, not in terms of the external world's responsibility: that of course would not have been permissible lest it compromised the very idea of independence.

The chapters have then one by one shown the exent to which there are major unresolved or worsening problems, as well as successes, over the review period.What caused population growth to explode so fast, towns to grow so fast, faster than the ability to provide basic services? Explanations of all kinds can be found in the preceding text: for some it is clearly a matter of declining death rates, itself the result of improved health care, and in time birth rates too will decline. To others, it is that Kenyans like having large families: children are the main point of life. Or as Hill (1991: 14) has put it: 'For years to come, the numbers of children borne by women will depend more on the social and exchange value attached to these individuals and less on the vigour of family planning or mother and child health programmes.' Whatever the cause, the magnitude of population increase is not in doubt. Average population growth rates of 3 per cent imply doubling times of less than twenty-four years. That implies half the population at any time are under the age of 24. And they need to be fed, educated and jobs need to be found.The dependency ratio may get worse if the awful spectre of AIDS confirms the dire predicitions of some.

The job of managing the development of the new states fell first on the shoulders of the new leaders and their governments. Their task has been unenviable, since the expectations placed on them have far exceeded the capacity of virtually all of them and a sense of urgency naturally underlay independence. Therein perhaps lies some of the responsibility for large and worthless aid projects, and should we not anyway have expected early failures as part of the learning process? To blame it all on outside experts is to say that African politicians have no responsibility at all, yet clearly they have had. And grandiose schemes did have a precedent: the image of the Soviet Union in the 1960s was of a society that had dragged itself from feudalism to superpower status by large-scale planning, avoiding external dependence. Perhaps the fault lies in the genuine but unrealistic hopes of all concerned for really rapid strides to be made, to change whole socities over night. The sense of urgency perhaps led to the wrong result: more haste less speed would have been better. But the need for haste is ever more obvious. Urgency, call it what one will, is needed at every level now even more than twenty-five years ago.

Even if many of these states have in some essence been 'artificial', they have survived as states even while the form of government has gone through cycles of dirigiste paternalism, despotism, collapse and reformation. As Clapham (1991: 19) has put it: 'The most basic reason in believing in the survival of the African state is that it is needed: no viable alternative to it has ever been suggested, and those countries where it has collapsed – Chad, Liberia, Somalia, Sudan, Uganda provide no stepping stone to the creation of any new social and poltical dispensation, but instead cry out for the restoration of the basic framework of order that only the state can provide.'

The state has survived, but at a cost. Sovereignty has had its limits exposed. An early move for many states was to dismantle colonial financial relations, to pursue independent policies. In many this leads to the chronic overvaluation we have noted, and the distortion of economies to induce protected high-price industries, and discrimination against agriculture. As feeding the populations has become increasingly difficult, and growth has failed to materialize, failure has forced many into the adoption of IMF restructuring programmes: thus dependency has been reintroduced, a new form of multilateral colonialism.

This new colonialism is not an answer either. Collier (1991: 20) suggests that 'Africa needs a new decolonization: donor conditionality should be replaced by the constraints of inter-governmental reciprocity'. Part of what he has in mind is the strengthening of African international agencies, be it common markets, or currency unions. He notes that it is timely to do so now, as Europe too is forming ever closer union. The power of Europe as a bloc has been mentioned many times above: in imposing tariffs and quotas on imports, on restricting guest worker immigration, and in dumping agricultural surplus. Co-operation within Africa is growing: it is ECOWAS that is sorting out Liberia's collapse, not directly the USA.

But further international co-operation within Africa and the

Middle East is desperately needed. War in the Horn of Africa has been a major factor in turning drought into famine, and war in the Middle East has wasted decades of investment in Iran, Iraq and Kuwait, not to mention the millions of lives. It has also had a major impact on the whole of southern Africa, and a resolution of the South Africa problem could do so much to transform the whole region.

Most of the industrialization policies have woefully underperformed. But there is one regional base – South Africa – which does have the infrastructure and the market to be a major stimulant to industrial activity in the region has a whole, if normal trade patterns can be restored, following political accommodation.

The combined effects of population growth, macro-economic policy, and political instability, have had major impacts on the internal geography of all of these states. The most obvious is in the phenomenal increase in urbanization – the rate in Africa is currently the world's highest (excluding some Middle Eastern states which have high levels of international immigration). The towns have grown at a rate which precludes the provision of basic sanitation, potable water and other services, let alone adequate housing, in all except the oil-rich states of the Middle East. Even semi-European middle-income Turkey has an appalling problem with overcrowded urban slums. The impact on agriculture of rural emigration has been harsh too and throughout much of Africa and those nations of the Middle East with a significant rural sector village populations are 'top-heavy', dominated by the elderly. The relative wealth of villages and the abundance of food there is a thing of the past. Instead, the elderly struggle to keep the family farm going and to provide necessary hospitality for migrant family members, and for friends.

The causes and consequences of environmental change are not clear. It is evident that there has been a twenty-year period of drier conditions in the Sahel, compared with the previous thirty years (Grove 1991), and this coupled with population growth and above all war, has as noted above caused famine. But there have been previous dry periods, and it is not possible to say either that the present one will continue, or that an equivalent one will not return. There is local evidence of land degradation – this is not a new phenomenon, but perhaps the scale is getting worse. It is ironic in a continent which in some ways is seen as under-populated, with populations too dispersed to permit 'modern' development. But attempts at villagization or the settlement of nomadic peoples has often contributed to local environmental degradation.

What always seems remarkable is that so little has been done to promote traditional farming. There is abundant evidence to show that traditional farmers are efficient at what they do and have the knowhow to manage difficult landscapes with which they are intimately familiar. There is plenty of room for improvement of traditional farming systems, but for too long Africa and the Middle Eastern nations trying to raise farm output have been reluctant to listen to their farmers whose low-tech methods were considered inferior to modern agriculture based on Western models. This attitude is at last changing, and the value of traditional farming methods are now being studied with the intention of developing them to improve agricultural output and solving the food crisis that exists in many states, particularly in Africa.

For rural and urban populations alike, and throughout almost the whole region with the exception perhaps of parts of Central Africa, water shortage is a major issue. It is a major issue between Turkey and downstream Syria and Iraq: there were hints water could have been used as a weapon against Iraq, with Turkey turning off supplies from its new dams. It is an ongoing saga in the Nile Basin, and in the Senegal Basin, where the rivers' waters are to be shared by neighbouring countries through water control projects. As the climate changes and the population grows, there is increasing pressure on water resources and increasing need to put water to the best use for all – without adding to environmental problems. Salinity from injudicious use of irrigation water has caused major problems in both the Middle East and Africa, and as aquifers become overused, the depth of the water table has fallen significantly in parts of the Middle East, North Africa and right across the Sahel. Certainly for the drier parts of the region, providing an adequate supply of water of acceptable quality is a major challenge for the future.

Many political leaders have pondered the problems of an African model for development. Part of the problem has been always that this has been seen in opposition to the Western model, not necessarily in terms of its own merit. Socialism with an African face as in Nyere's Tanzania had a chequered history, but now internationally that model is being abandoned. The same quest for an alternative is felt throughout much of the Middle East, and it should not be underrated. But here there is a force which does provide some alternative thinking – Islam. Politics in the Middle East in particular have been greatly influenced by Islam and the Islamic revival that occurred in Iran in the early 1980s has intensified the impact of Islam on politics. Perhaps there are then two models emerging: the Islamic state of the north, and a new form of multi-party democratic state in the south – but whose advent and success should not be prematurely assumed.

While there has been so much change in Africa and the Middle East in political and economic terms, poverty has endured. Essentially, Africa's non-oil economies have been unable to adjust swiftly to changes in world economic conditions. Their heavy dependence on the production of low value, primary produce has left them uncompetitive and largely dependent on aid which now forms a significant and permanent part of the capital account of many African nations. Even where, as in Iraq, Iran and Saudi Arabia, oil has generated wealth, there are many who have not benefited: the distribution of wealth internally is as difficult as its distribution or redistribution internationally:

> the main danger to African states ... would come from ... the outside world's indifference: the danger that, tired of propping up states which cannot maintain themselves (and even of feeding people who cannot feed themselves), directing their capital to more efficiently managed economies offering higher rates of return, and freed from the need to compete diplomatically with a defunct communist bloc, the Western world might simply abandon Africa to its own inadequate devices.
>
> (Clapham 1991: 16)

The West is patently unable to leave the Middle East to its own devices, for reasons of self-interest. Perhaps in both cases what is needed is self-disinterested involvement. It is needed urgently. Neglected change has a habit of catching up on the uninvolved.

REFERENCES

Brookfield, H. (1975) *Interdependent Development*, London: Methuen.
Clapham, C. (1991) 'The African State', paper presented to the Royal Africa Society Conference on *Sub-Saharan Africa: the Record and Outlook*, St John's College, Cambridge, April.
Collier, P. (1991) 'Africa's external relations:1960–1990', paper presented to the Royal Africa Society Conference on *Sub-Saharan Africa: the Record and Outlook*, St John's College, Cambridge, April.

Grove, A.T. (1991) 'The African environment', paper presented to the Royal Africa Society Conference on *Sub-Saharan Africa: the Record and Outlook*, St John's College, Cambridge, April.

Hill, A.G. (1991) 'African demographic regimes – past and present', paper presented to the Royal Africa Society Conference on *Sub-Saharan Africa: the Record and Outlook*, St John's College, Cambridge, April.

INDEX